PRACTICAL WORK

IN UNDERGRADUATE SCIENCE

ADVISORY COMMITTEE

Dr M Gavin (chairman), formerly Principal, Chelsea College

Professor A Becher, University of Sussex

Professor R J Blin-Stoyle, University of Sussex
(succeeded by Professor J P Elliot)

Professor W E Burcham, University of Birmingham
(succeeded by Professor W F Vinen)

Professor E J Burge, Chelsea College

Dr C C Butler, Director, The Nuffield Foundation
(succeeded by John Maddox)

Professor R G Chambers, University of Bristol

Professor M J Frazer, University of East Anglia

Professor J B Hasted, Birkbeck College

Professor D W O Heddle, Royal Holloway College

Professor K W Keohane, Centre for Science Education,
Chelsea College

Dr M R Parlett, National Foundation for Educational Research

Professor F R Stannard, The Open University

Professor C A Taylor, University College of South Wales

Dr R Thorburn, Liverpool Polytechnic

NOTES ON CONTRIBUTORS

Roy Davies teaches physics and electronics at Royal Holloway
College.

Martin Harrap taught physics in schools, and has worked for the
Nuffield Advanced Physics Project, and also for UNESCO.

Susan Kay teaches physics at Royal Holloway College.

Roy Lawrence is in the Physics Department at Liverpool
Polytechnic.

Sid O'Connell teaches in the Institute for Educational Technology,
University of Surrey.

Jon Ogborn, project coordinator and editor of this volume, is
Senior Research Fellow at Chelsea Centre for Science Education.

Patrick Squire teaches physics at the University of Bath.

Robert Whitworth teaches physics in the University of Birmingham.

Practical Work in Undergraduate Science

Contributors

E.R. DAVIES

MARTIN HARRAP

SUSAN M. KAY

ROY LAWRENCE

SID O'CONNELL

JON OGBORN

P.T. SQUIRE

R.W. WHITWORTH

in association with

P.J. ASPDEN

P.J. BLACK

WILL BRIDGE

NEIL B. CRYER

J.A. EADES

R. EARDLEY

J. FENDLEY

BARBARA HODGSON

S.J. PENTON

Editor and series editor

JON OGBORN

Published for
The Nuffield Foundation by
Heinemann Educational Books
London

Heinemann Educational Books Ltd

London Edinburgh Melbourne Auckland
Toronto Hong Kong Singapore Kuala Lumpur
New Delhi Ibadan Nairobi Johannesburg
Lusaka Kingston

ISBN 0 435 69582 7

© The Nuffield Foundation 1977

First published 1977

Published by Heinemann Educational Books Ltd
48 Charles Street, London W1X 8AH

Printed in Great Britain by
Biddles Ltd., Martyr Road,
Guildford, Surrey.

General preface

The Higher Education Learning Project is a working alliance of teachers in higher education, initially of physicists. We began in 1972 with a group drawn from the Universities of Birmingham, Surrey and Sussex; Birkbeck College, Chelsea College and Royal Holloway College in the University of London; and from Liverpool Polytechnic. Over the next four years we were joined by teachers from a considerable number of other Universities and Colleges.

The activities of the project fell into four areas:

Individual Study: experimenting with, and looking into the value of methods of teaching which placed less reliance on the lecture.

Tutorial Teaching: investigating the problems of small group teaching in science, and trying out new materials for tutors to use.

Laboratory Teaching: investigating the problems of laboratory work, and the advantages and difficulties of a variety of kinds of innovation.

Motivation: a large scale study, interviewing students at many universities, with a view to understanding better the problems of learning as they see them.

Many departments of physics helped in the work of the project. We are grateful to the Department of Physics, University of Birmingham, for providing facilities for Barbara Hodgson; to Chelsea College for facilities for the co-ordinator, the project secretary, Martin Harrap and Dietrich Brandt; to the Institute for Educational Technology, University of Surrey, for providing facilities for Will Bridge and for releasing Sid O'Connell part-time; to the School of Mathematical and Physical Sciences, University of Sussex, for releasing Peter Unsworth part time; and to the Physics Department, Liverpool Polytechnic, for releasing Roy Lawrence part time.

Much of the work of the project has been done by people who

gave freely of their time, without reward. The project is particularly grateful to Lewis Elton, who organised the individual study activities of the project, and to Joan Bliss who developed the motivation interview, trained physicists in interviewing, and organised the whole study.

The project owes its thanks to the very many teachers, in a large number of departments, who involved themselves in the work. Their names appear in the publications with which they were particularly associated. We are also grateful to the many students who have talked to us, and who have been at the receiving end of various innovations.

We also wish to thank the Director and Trustees of the Nuffield Foundation for their support; and the members of the Advisory Committee, notably the chairman Dr Gavin, for their continued advice and help.

Finally, everyone concerned in the project is in debt to Paul Black. He played a major role in initiating the project, acted in all but name as joint co-ordinator, and gave generously of his time, energy, and insight to every aspect of its work.

We began as a group of physicists, but it was never our intention to concern ourselves solely with problems of teaching physics. We have had useful discussions with teachers in a number of other scientific disciplines, and this experience is the basis for the belief that these books will interest many besides physicists. Nothing would please us more than to have made some contribution to the discussion of teaching problems in the academic community at large.

Jon Ogborn
Co-ordinator

Preface to this volume

It will not surprise the reader that laboratory work was the first problem area mentioned when, at the start of the project, we asked ourselves what we were most concerned about. Nor will it surprise him to be told that we found it less tractable than others.

The work, of which this book is one outcome, had two phases. In the first, ideas and materials were exchanged between staff in various departments, out of which grew innovations and new initiatives in some. The project is grateful to Dr Dietrich Brandt for the role he played in this phase.

The second phase grew out of the realisation that the problem was not to find or invent new ideas, but to understand better the consequences of various sorts of change. Few were wholly satisfied by what they had, whether conventional or novel, so it seemed that a rather deeper study of the working of laboratories of various kinds in different settings was needed.

Accordingly, a team of physicists within the project spent a year making visits to eight physics departments, looking at four first and four second year laboratories. In parallel, Roy Lawrence made visits to another eight institutions to look mainly at third year project work. We collected materials, observed students and staff at work, and interviewed them. The book is based largely on these visits. We are indebted, for their willingness to help and to talk to us to staff and students at:

Bath	Manchester
Birmingham	Nottingham
Bristol	Royal Holloway College
East Anglia	Sussex
Lancaster	UMIST
Liverpool	University College, London

The team which made the visits was:

Penny Aspden	Martin Harrap
Paul Black	Barbara Hodgson
Neil Cryer	Susan Kay
Alwyn Eades	Jon Ogborn
Jack Fendley	Patrick Squire

A group consisting of the above, together with

Will Bridge	Roy Lawrence
Sid O'Connell	Simon Penton
Roy Davies	Robert Whitworth
Bob Eardley	

planned the book and the laboratory investigations for it, and criticised and improved drafts of material for the book. It is to their collective judgement that the book owes what value it may have for staff engaged in the difficult job of laboratory teaching.

While all parts of the book were discussed and criticised in this way, so that no one person would wish to claim sole responsibility for any part of it, the main responsibilities for the final versions of chapters were:

chapters 1, 3, 4, 9	Jon Ogborn
chapters 2, 5	Martin Harrap
chapter 6	Roy Davies, Sid O'Connell, Jon Ogborn, Patrick Squire, Robert Whitworth
chapter 7	Roy Lawrence
chapter 8	Martin Harrap, Susan Kay, Sid O'Connell, Jon Ogborn, Robert Whitworth

with Paul Black assisting the editor in preparing the final version of the whole book.

Finally, we wish warmly to thank the many other university teachers who, over the years, attended meetings, shared their ideas and worries with us, and welcomed us into their laboratories. Theirs has been a vital influence on the development of the book.

Jon Ogborn

Contents

1. What this book is about

1.1 UNDERSTANDING THE PROBLEMS OF PRACTICAL WORK

This book is about understanding the problems of practical work. It is addressed to a variety of people: to members of staff or post-graduate demonstrators whose concern is simply to try to do a little better in day to day laboratory teaching; to people who run existing laboratories or are designing or reshaping a teaching laboratory; and to those who are concerned, in departmental discussions and committees, with arguments about the role and nature of practical work.

Not every reader will find all parts of the book equally useful or acceptable. Indeed, not everyone will accept that all the problems discussed here are problems at all. The function of this first chapter is to indicate the content and character of the various parts of the book, putting a variety of issues before the reader, so that he can see the scope of the things which have been taken to be worth discussing, and to indicate where that discussion can be found.

We have tried, throughout the book, to provide a critical discussion of issues, based on evidence and examples. The evidence and examples were collected in an extensive series of visits to teaching laboratories by a team of physicists. This is not to say that the book purports to be 'scientific' in its approach: to attempt to measure a laboratory and the things that go on in it would be about as sensible as to attempt to measure a marriage. But we have attempted to be dispassionate; not to bend evidence to arguments it cannot support. Even so, the relation of evidence to issues and conclusions remains problematic, and the final section of the chapter outlines our approach to resolving these difficulties so far as we were able to.

The book has been written at an interesting period in the development of practical work; a period in which there is much questioning of its value, a good deal of experimenting with new methods and patterns of organisation, but not any clear and agreed way forward.

1

Somewhere over all current thinking, however radical, falls the shadow of traditional university practical work. It is a great tradition, of which Searle was one of the best but by no means the only exponent. The traditional laboratory contains a number of experiments, many of them ingenious and elegant, each in some way a microcosm of the art of experimentation, though the emphasis may be laid more on one aspect than another. Each is more or less equivalent to the others, in the thinking it demands, the time it takes, and the difficulties it presents. Usually, students do the experiments in no special order; a pattern called in this book a 'circus'.

The following discussion* with a student at a large, well-respected university, said by staff to be 'serious and straight-forward - a possible first', may help to bring the foregoing neutral account to life.

'I don't particularly enjoy laboratory classes at the moment. They are OK, but... the experiments rarely work, so there is no real sense of satisfaction. I have sort of done every natural constant in the Universe this year... tried to determine it.'

'... tell me a bit about the lab classes. How were they run for you?'

'You work in pairs, and four pairs have a post-graduate demonstrator over them... Over about each... four or five demonstrators there is a lecturer type. He is OK really, but the demonstrators are a pretty friendly lot. They give you this massive wad of printed notes. Sometimes it tells you what to do, and sometimes it doesn't. Sometimes when it does it is a load of rubbish...'

'... a mass of notes?'

'At the beginning of the year you get, oh about fifteen sheets, with fifteen or so experiments you do in the year. The idea is that you plough through these - they eventually go into theory you haven't done at all, so it is pretty meaningless. Every now and again, about four times a year, we have to do this report of an experiment - a full write-up of everything that happened, which means that you copy all this theory out, even if you don't understand it, and that seems a bit pointless... it is tedious... because you waste about six hours when you could be doing other work. The idea, I think, is that it is meant to be good training for research or something, but it seems very pointless at the moment.'

'What sort of experiments do you do?'

'... the ones that work are a sort of mechanics type - where you swing pendulums around. There are a fair number of heat ones, which I always manage to treble the value on... The electricity type aren't too bad, in fact they are quite enjoyable really - I don't know why.'

'What do you mean by "not working"?'

'Well, you do the experiment, and everything seems to have gone perfectly. Then you go and start plotting graphs, and straight lines are curves, and it just doesn't work out right - I don't know what causes it. But the experiments that do work, say when you sit there swinging a pendulum and get a nice straight line off it are incredibly boring.

In fact all the easy experiments, whether or not they give good results, are the boring ones, because you just set something up and then... take about six sets of readings off it - you can either swing a pendulum at six different lengths, or heat a chamber up to six different temperatures, waiting ten minutes between each temperature. You are just sitting there most of the time.'

'You used several words: "meaningless" and "pointless" ... How does it feel to be doing something you think of like that?'

'Well it is a drag, really. I mean these reports are a nuisance. The first one I did took about nine hours, because I spent about three of those hours just reading through a book on how to write physics reports - deciding what I was supposed to be doing.

The ones after that I have done in about six hours - but essentially you just copy out the theory and all that from the printed sheet, draw a diagram, write a method, and tabulate data. Then you come to the errors, which is the really interesting bit. I mean, you get six points on a graph... which gives you a possible combination of fifteen pairs of points, and you find the gradient between each of these pairs - you find fifteen gradients... and

. . . .

*Interview conducted in the course of the project's motivation study. See the companion volume, Students' Reactions to Undergraduate Science. Other quotations here are from the same study.

average them, and take the deviation from the average
... It's just arithmetic, but it's tedious.'

'How do you feel when you get up in the morning on the
day for laboratory work?'

' "Oh, God, not that again". If a thing is going well...
it is OK, but when things start going wrong, I get fed
up with it. I tend to swear at voltmeters and such...
Sometimes when I go home afterwards, I think-that
wasn't too bad - in fact it was quite enjoyable, but I
never look forward to it.'

Nobody would say that this was the whole truth, but it is a
part of the truth, and a part missed by general discussion of
the aims of practical work. One aim of the book is to try to
bridge the gap between what happens as individuals see it, and
what is meant to happen. Part of that task is to try to understand why things come out as they do, for better or worse, so
as to have some basis for arranging things so that what is
worthwhile happens more often. And that involves deciding what
is worthwhile. Perhaps the following can stand as an example:

'I could choose the experiment, knowing that I didn't
understand it. And the challenge, of being able to walk
up to something, and not know a thing about it, read the
small blurb they provide for you, go and get the books
and find out about it. And actually succeed, and come
out with an answer that's almost right. That's what I
think physics is about... To actually have a problem
and solve it, and actually achieve something. Other
people know what the latent heat of nitrogen is, much
more accurately, but I feel satisfied that I've been able
to work it out all by myself... instead of an endless
round of learning, of putting facts into my head. Something that I can actually do - it seems almost useful.'

1.2   STRUCTURE OF THE BOOK

The book looks at laboratory work from a variety of points of
view. First, it attempts to get some grip on reality, by
describing what happens in various laboratories, drawing
contrasts and comparisons. Then various chapters explore
possibilities for change, great and small, ambitious and modest,
with a commentary relating them to the general framework of
practical work, and indicating where possible important aspects
of how they turn out in practice. Finally, the last chapter

confronts the question of the various aims of laboratories, and of what light the ideas in the rest of the book cast on the problems of achieving them.

1.2.1 PEOPLE

Chapter 2 concerns people in laboratories; their roles and the circumstances that affect how they play those roles.

Sometimes, by chance or by design, staff and students in the laboratory go through the kind of experience most would wish for:

> '... I can't remember his exact words, but he looked at it and sort of said, "You know something, that's right. It works - it's bound to - come here, Derek, have a look at this - he's done it." It was probably the first time anyone had done that bit of the practical ... it was supposed to be theoretical - just design the circuit, rather than construct it.'

Sometimes, however, perhaps because of the nature of the people or of the events, things are not so good:

> 'I know when I felt completely lost, and that's in the labs here... We were given a piece of paper telling us what to do, and then they just put us in front of the apparatus and said, "Carry on"... and we'd no idea how to start... they had three demonstrators between about thirty of us... and they just came around every half an hour and asked us how we were getting on.'

Such events, good and bad, show that there is a need to think out the nature of the job, as it presents itself to staff, post-graduate demonstrator, student, and technician. Questions about why change is difficult, or about why students are sometimes nervous and sometimes over-confident need asking. Central to the whole chapter is the paradox that the laboratory is at once clearly one of the most central and valuable parts of any science course, and at the same time the one which raises most doubts and the place few admit to enjoying teaching in.

1.2.2 PROCESSES

Chapter 3 looks at a number of things that go on in laboratories;

at what happens when students do experiments; at what demonstrators do; at how work is marked; and at the overall question
of how staff and students learn what is expected of them through
overt and covert channels of communication.

Several of these issues are illustrated by the following story
about getting work marked·

> 'And then they expected a fantastic write-up. This was
> the first write-up we'd done, and I think my mark was
> $3\frac{1}{2}$ out of 10... I thought, that's bad, if that's all I'm
> going to get... I'm not going to do very well... I thought
> I deserved more. I did try for that write-up - I didn't
> just do it in half an hour, I put quite a bit of work into
> it. I was sort of wondering whether or not to complain,
> and I thought, well, I haven't really got any right to
> complain at all... I mean, if it's not what they want,
> it's not what they want. You can't have marks for what
> I think is right.'

The central activity, however, is doing experiments. What,
it needs to be asked, is success and how is it to be encouraged?
How is time actually spent, and how much of it is productive,
and in what sense? The answers are sometimes simple but not
obvious:

> 'That was marvellous... The experiment worked out
> wrongly, but why it worked out wrongly really made me
> feel good... To start with I felt absolute surprise.
> But then after accepting that it had happened, it was a
> feeling that you have really learnt something. It was
> learnt in such a way that you'd never forget it.'

1.2.3 PARALLELS

It must not be supposed that traditional practical work dominates
the laboratory scene, with only isolated pockets of innovation.
There are a considerable number of published accounts of
laboratories which depart from tradition in a number of respects.

Some have adapted ideas from programmed learning, laying
stress on isolating rather precise objectives and devising
exercises intended to achieve them. Others have tried to make
practical work more 'open-ended' - a term which means as
many things as it has exponents. Yet others have tried reorganising the laboratory, working perhaps in groups instead of
with students doing isolated tasks singly or in pairs.

The differences between these ideas, or between them and the traditional form, ought not to be overstressed. Chapter 4 attempts to discuss them within a general framework which relates them, and clarifies so far as possible some of the terms people use. What, for example, is 'open-endedness'? Is the following an instance, or is its message more one about flexibility?

> '... nobody had done it before, it was just a bit different, and I spent almost the whole afternoon playing about with it... Most experiments are set, and it's all very standard. I think it was one of the supervisors walking around, said, "Oh, that's interesting", and I said, "Well, look, I'm not getting much out of this. Can I look into it?". So I did... It wasn't of any particular interest, but I'd done it, and I was pleased and happy. I really thought I'd got something out of it. All right, it wasn't one of the set experiments, and it didn't help me get experience with all the different equipment, but I'd done it and it was different, and I enjoyed it.'

However, there is perhaps more value than analysis can offer, simply in seeing possibilities others have tried. One may or may not like them, but there is much to be gained by thinking out one's own reaction to them.

1.2.4 PLACES

Chapter 5 describes in some detail the form and arrangement of practical work, as we observed it in action in a number of contrasting laboratories.

The accounts are selected so as to bring out points of special interest to those responsible for laboratories. They cover some well designed and well managed traditional laboratories, as well as others which depart from tradition.

Everyone knows that each place has its own unique character; indeed one thing that is now very clear to us is that transferring an idea from one place to another can, and usually does, transform it in important ways. Thus a main purpose of this chapter is to show how various laboratories form self-consistent wholes, and how the way the features that they have fit together determines much of what can happen.

1.2.5 PROGRESS REPORTS

In chapters 6 and 8 we have collected a number of progress reports from people who have introduced significant innovations of various kinds.

There is value in seeing how they view their own ideas after the experience of trying them out, so the reports are mainly written by those involved. At the same time, there is value in having an outside view also, so we have taken advantage of opportunities to visit and observe these laboratories, and out of that to write a commentary which both offers some evidence of how they work, and relates them to one another.

The reports describe very varied innovations. Some lean towards programmed learning, for example, and sometimes find that it is not always as simple as it seems:

> '... it occurred to me that it's all very well having programmes, but you have to be very careful about how you word them, because I was confused a number of times... there was one about voltmeters, and how they worked... I was very annoyed because I spent half an hour trying to work it out - I knew it was fairly simple so I was annoyed at not being able to get it out. Then we went and asked the guy who was doing it what was meant, and he said, "Oh, but there is so and so", you know; if I'd known that I could have got it out in a couple of minutes.'

Others are concerned with new patterns of organisation, which might help by giving staff more direct and personal responsibility. A particular example of one outcome is given. Yet others want to look at the whole content of the laboratory, asking why all experiments should be equivalent and serve broadly the same ends. Can one not be rather more flexible and eclectic?

> 'I think the tedium was broken by being able to get up and have a look at little experiments - that didn't really prove a lot or anything - but I think in your own mind you got a clearer picture... It showed me that things in the physics course weren't just things out of a text book... You can see it happening, you know that it really exists the way they say it does.'

1.2.6 PROJECTS

> '... you're not watched over, you're not confined. You can do the project they way you want, and if you make a mistake, you take the responsibility for it. You get the credit when you do it right, and you get moaned at when you get it wrong, but at least you know, "That was all mine".'

The quotation illustrates one reason why students like projects, and so why they have often replaced other practical work in the final year. Chapter 7 is based on a study of project work which tried to go behind such rather general attitudes, to find out more about what 'freedom', 'reality', and other such terms meant in practice.

The chapter explores what supervisors of projects do, and asks how projects might be organised so as to increase the benefits they offer.

As in the rest of the book, no simple prescriptions can be offered, and nor can such a complex and variable phenomenon be studied as if it were a predictable machine. Rather, what can be done is to attempt to clarify arguments, especially by seeing what they mean in practice, and out of that to present possibilities and alternatives for consideration.

1.2.7 PURPOSES AND PLANNING

The remarks above apply even more strongly to the final chapter, which tries to draw together the various threads of the book, in a discussion of the aims of practical work and the influences which aid or hinder their realisation.

The arguments in this part of the book depend on its four underlying aims. The first aim is to clarify discussion, in part by suggesting a framework of terms of discussion, and - more important - to link the discussion as much as possible to the real, actual, and practicable.

The second aim is to expose and consider a wide variety of possibilities, so that discussion of what to do is as widely informed as it can be.

The third aim is to bring out - or at least, to invite the reader to do so for himself - some general conclusions. Nobody will suppose that these conclusions will all be secure, or that they will not be in any conflict with one another. As in ordinary life, so in the laboratory, well planned events can have bad or unexpected consequences, and ill-conceived ones can have remarkably good byproducts. But this need not mean that the picture ought to be painted a uniform grey; rather, there is a complex pattern of light and shade, with shafts of light and pools of darkness in unexpected places. Thus the fourth aim is in some sense to complicate the third, by showing how, although some conclusions can be argued for, they cannot be applied sweepingly or without discrimination.

Some examples of the kind of conclusions which might be argued for are:

What happens by no means always matches what is supposed to be happening.

There are all sorts of critical phenomena in laboratories: seemingly small influences which largely determine important aspects of what happens, or how people see events.

Notwithstanding the last point, laboratories are in many ways enormously stable, with changes making much less difference than might be supposed or was intended.

Who the people are, what they are like, and how they behave, makes a big difference, perhaps more important than any planned organisation.

The mythology of the laboratory - that it is a place where one learns experimenting by direct contact with those skilled in experimental research - needs to be called in question. There is less contact than imagined, and it is not often of this kind. But it can happen, and deserves recognition.

The student is often in a situation in which he switches backwards and forwards between doing physics, and keeping teachers satisfied. The effect is inevitable, and not necessarily to be regretted, but is important in understanding what happens.

The laboratory time scale is long. What happens in one experiment is rarely enough plausibly to support the hopes built on it.

Almost any pattern can produce good, or bad outcomes, or both. Amongst patterns, the traditional set experiment laboratory has a robustness which deserves respect. But innovations can introduce a sense of purpose which makes a considerable difference.

Students differ enough for it to be reasonable to try to match practical work to individual needs.

Staff differ enough for it to be implausible that any one pattern can be operated by anybody. They do best when they do what they are individually good at.

True remarks about any laboratory tend to be blindingly obvious, once stated. But from the inside, it is not always so easy to see their truth, or their consequences in action. Learning to see what is going on can be important.

Such ideas are at best thought-provoking. But, in a time when practical work is being called in question, and ideas are fluid without yet taking any very clear direction, that may be the best thing to offer. At least one may hope to move a little beyond the position from which a member of staff sadly said to us:

> 'You have total authority in running the lab. Just as long as the students keep coming, are processed, and get marks that are not too bad, you are left alone. Nobody wants to know.'

1.3 OBSERVATION, EVIDENCE, AND ARGUMENT

The basis of the book is observation and argument. It has in part come out of visits by a team of ten physicists to eight different teaching laboratories, both first and second year; visits by one to project laboratories; several briefer visits and, inevitably, recollections of previous experience.

We made first hand observations of students and staff at work; talked to both about what they thought they were doing; and collected together and argued about our different impressions of individual places and of comparisons that could be drawn between them. In so far as these reflections represent the views of physicists interested in laboratory work, we hope that they will interest others who have similar concerns.

Clearly such a study is unlike the scientific investigation of, say, a range of materials to be used in various applications, tempting though it might be to try to see what ought to be done in that kind of way. We think, on the contrary, that a parallel with scientific investigation is misleading and damaging, even as an aspiration. It is a peculiar difficulty for scientists involved in education that they naturally think in scientific metaphors; of 'factors which affect' - of 'parameters' - and so on. The problems of education, or so it seems to us, are not appropriately conceived in this way. They owe something to facts, but the facts are ones which shift, even reverse, their meaning

when looked at from different points of view. They need dis-
passion and an openness to contrary arguments, but the dis-
passion is of a kind that requires, not detachment from any
special point of view, but a clear statement of predisposition
or even prejudice.

More telling, perhaps, is a commonsense argument. Few
teachers would be satisfied with a general written account of a
laboratory, and none would feel able to put it into practice on
that basis. Rightly, they prefer to go and see for themselves.
One reason is that one can trust more what a person visibly
does than what he says he does, however sincerely he says it.
Another is that insight comes, not from a mass of factual
detail, but from particular instances which, despite their
accidental quality, illuminate by significant detail what before
was not understandable, or lacked vividness and force.

In a book such as this, which uses concrete detail where
possible so as to bring into more effective relation general
thoughts and the actuality of what goes on, the very selection
of evidence and of the points of view from which that evidence
is seen, are inevitably distorting and potentially dishonest.

The distortion cannot be avoided. There are no neutral
points of view. Our only protection has been to work together
as a group, trying hard to minimise the effects of the prejudices
of individuals, by testing out interpretations against the experience of the others. Even so, all we can offer is the question,
which also seems to us to be the proper one to put:

'This is what we think, and this is why.
What do you think?'

2. People

2.1 DIFFERENT SORTS OF PEOPLE

The people in university undergraduate teaching laboratories fall into obvious categories: those whose business it is to teach; those whose business it is to learn; those who give technical help; and, occasionally, visitors.

Of those who teach, some are senior members of staff with experience of successful research. One or more of them has responsibility for running the laboratory. Others, and in a large department the majority, 'do their stint' without contributing to the direction of policy. Those who teach also include research students, of an age closer to that of the undergraduates they teach than to that of the staff they assist.

Undergraduates outnumber all other groups, but their connection with the laboratory is the most transient. Staff divide them mentally into 'good' and 'weak' ones, while being unsure in some instances to which category an individual belongs.

Technicians are a small group, but not infrequently have individually the longest association with the laboratory. Their often considerable influence is rarely obtrusive.

Visitors play no part in the normal functioning of the laboratory, and would have no place in this account if the book could have been written without them. But their possible distorting influence on whatever they may have observed and recorded must not be forgotten.

2.2 WHAT THE JOB IS

The people are familiar, and the job may seem prosaic, but the tradition to which they belong is grand. Since Bacon wrote,

> 'The secrets of nature betray themselves more readily when tomented by art than when left to their own course'

experiment has proved to be one of the main ways, and certainly a crucial way, in which what we call science has grown, and has carved out some intellectual territory in which superstition has been eroded away. It was William Thomson who, charged with the task of teaching Natural Philosophy in the University of Glasgow, first seems to have introduced experimental work for students. Speaking in retrospect of this innovation*, made in 1846, he explained how it came about:

> 'I had occasion to undertake some investigations of electrodynamic properties of matter, to answer questions ...which could only be answered by direct experiment. The labour of observing proved heavy; much of it could scarcely be carried on without two or more persons working together. I therefore invited students to aid in the work...Soon after, other students, hearing that their class-fellows had got experimental work to do, came to me and volunteered to assist in the investigation. I could not give them all work in the particular investigation which I had commenced...but I did all in my power to find work for them on allied subjects.'

By 1863, there was a laboratory for 'Experimental exercises and investigations' open daily from 9 a.m. to 4 p.m. Clearly, Thomson spoke of 'experiment' as of a trial whose outcome cannot be predicted, and it is experiment in this sense, not in the sense of practical work set to be done in the laboratory, which ultimately it is the purpose of the university teaching laboratory to impart. It is the central place of experimental work in science which makes the teaching laboratory the single most essential part of any science course; the one part without which a course cannot properly be described as science.

The student in the teaching laboratory is far from the frontiers of knowledge. The results of his 'experiments' are normally unpredictable mainly in their imprecision. But because teachers do research as well as teach, the student has always, if he can make use of it, access to genuine frontiersmen. Whether he can or does make use of it varies a good deal from student to student and teacher to teacher, but may in part be what distinguishes the 'good' from the 'weak'.

....
*Address at the opening of physics and chemistry laboratories, University College of North Wales, February 1885.

14

2.3   THE STAFF DEMONSTRATOR

In the university laboratory, the teacher's qualification is in
Bacon's art of tormenting nature's secrets, not primarily in the
art of passing secrets on to other people. As with secrets
generally, it is assumed that people will be anxious enough to
learn them for the wiles of the publicist to be superfluous.

In extreme cases such a lack of concern for artful persuasion
can be unsatisfactory. A minority of scientists has found in
research an occupation which, absorbing all their attention,
spares them unwelcome contact with other people. Since Ampere
and Cavendish might be thought to be amongst their number, it
is difficult to scorn their remoteness, despite H.G. Wells's
account of its effect on him when evidenced by another of the
same temperament:

> 'In those days I thought him one of the worst teachers
> who has ever turned his back on a restive audience,
> galloped through an hour of talk and bolted back to the
> apparatus in his private room.'

At the other extreme might stand Davy, of whom Coleridge
said that if he had not been the first chemist of the age, he
would have been the first poet. Davy may have had more skill
than a laboratory demonstrator absolutely needs, but any excess
cannot be thought counterproductive when Faraday was his student.

2.3.1   WHAT THE STAFF CONTRIBUTE

The staff demonstrators we met were well between any such
extremes. It seemed to be accepted generally that there exists
a teaching contribution which all can make, part of whose nature
is that it does not relate to any particular laboratory, to any
special branch of the subject, or to any particular experiments.
Rather it consists of a set of standard and quasi-eternal skills,
which once acquired, can always be applied no matter what the
task. The skills are not mechanical or manipulative; they are
analytical and critical.

Any experimental task, for which it is possible to think of a
method, presents alternatives, and the researcher compares the
alternatives critically and chooses the best which time and other
resources permit. The best very often means the most accurate,
so that speed and sureness in estimating precision are valuable.
The teaching contribution such a person can make is to be himself; to display those qualities and when appropriate to expound

their principles.

So, for example, if a student asks what resistor to use, it is not seen to be the demonstrator's duty to be able to point to one, but he should be able to carry the student through an enquiry as to what the resistor is to do, in what circumstances, and as to what would happen if it did nearly but not quite what was required of it, and so on.

Thus it is that the demonstrator is, so far as his special qualities are concerned, to a great extent regardable as a 'plug-in' component. His contribution would not be much affected by unplugging him from one laboratory, and plugging him into another. It is not even clear whether his value might not become greater for a time in this event, for it is sometimes said that he can display his qualities better still when he does not happen to know the trivial answers to questions, but is driven to arrive at them by taking visible thought.

The notion of the 'plug-in' demonstrator is a convenient one for universities, in which staff have calls on their time which are sometimes unpredictable, and often not less important or interesting than teaching. The ability to move demonstrators from laboratory to laboratory inhibits boredom and staleness, and is organisationally convenient in a system in which changing work-loads have often to be accommodated. It also makes it possible for a man to attend a vital committee, or make an urgent visit to a research laboratory, feeling confident that his temporary replacement will function just as well as he would have done.

It must, however, be a part of this concept that knowledge of students and of what they have been, are, and will be doing is not essential. Indeed, laboratories divide sharply between those in which this principle is taken for granted, and those in which it is not. Each kind is organised in such a way that the premise it assumes is made more or less true. Many other factors influence the style of experiments, the provision of scripts, the arrangements for giving advice, and the system for generating marks, but the way these and other elements cohere into a system cannot be understood without paying some attention to this dimension.

2.3.2   ACTIVE OR PASSIVE COMPONENTS?

A time and motion study of how the average staff demonstrator spends a six hour day in a teaching laboratory might reveal some such result as that he spends a total of three hours discussing and assessing students' finished work in confrontations lasting anywhere between ten minutes and an hour; a total of two hours split into fragments of between one and ten minutes helping students at the bench; and the remaining hour split into fragments which may last only for a moment but may rise to thirty minutes, sitting, standing, or walking slowly about doing nothing but keeping a general eye on things.

The job is one of which few staff demonstrators spoke with enthusiasm. It perhaps consists too much of reacting to what other people do or ask, and too often has occasions when initiative has to be secretly suppressed because it would be unwelcome or undesirable, for the job to be a satisfying one. This is not to say that it is not well done; indeed the tact and skill of many demonstrators, even in the art of doing nothing when that will most encourage learning, is impressive. Students too will often speak well of the way they do their job:

> 'The staff are very good. If you can't see where a wire goes you feel an awful fool, but they never say, "You idiot".'

Those who direct laboratories seldom speak of initiative in the way of planning and updating from their colleagues. One, asked if he received such help, replied that it was like getting blood out of a stone. Others, unasked, said much the same (but note in contrast the discussion of the Liverpool first year laboratory, chapter 8, section 8.3). Non-contributing staff demonstrators confirmed that all the credit belonged to the directors in question, the lonely nature of their task being the natural order, and a burden that each would no doubt shoulder manfully when his turn came.

One reason for this lack of general bustle may be the prevalence of the view that the old ways are best; that the principles of physical investigation were as well understood by Kelvin as they are now, and that it is these which are the proper study of a student, not techniques ten years old which may be discarded before he has occasion to use them. A good piece of teaching has therefore little built in obsolescence.

Another reason, commonly advanced, is that because professional advancement depends mainly on research, enthusiasm for teaching is quixotic. One experienced technician made a convincing case for the direction of a laboratory to be in the hands of a man old enough to be no longer anxious to make his reputation in the world.

17

However, beyond these simply stated reasons, there seem to lie others less easy to express, except in a general way by saying that the job is bigger than it looks. The next section attempts to size it up.

2.3.3 CHANGES IN LABORATORIES

A teaching laboratory may contain some thirty different exercises, each of which has apparatus to occupy a student for perhaps twelve hours. Perhaps an experienced demonstrator could scarcely set up such a new exercise in less than ten hours, including assembling apparatus, getting parts built by technicians, trying it out, writing the script, and so on. Many we saw had clearly had much longer times spent on them, but even taking this modest figure, the laboratory represents the investment of at least 300 hours of demonstrators' time.

Further, circumstances require that each experiment must fulfill a number of quite arbitrary conditions, such as taking roughly equal times and being of roughly equal difficulty. Any attempt to vary these characteristics for one or two means that the collection as a whole loses organisational coherence, so that any but minor variations involve modifying all the exercises and an outlay of the order of 300 hours. Even the keenest demonstrator cannot contemplate this, so wholesale change is rare.

Replacement of one exercise by another equivalent to it in these respects is a reasonable solo performance. It often appears desirable, and is often done, in cases where there exists a new technique which is not represented in the laboratory, or in cases where a phenomenon which has caught the demonstrator's interest can be shown because of some new development in hardware. So a demonstrator might find it easy to assemble an exercise on ferro-electric crystals, and use it to replace one in which a mirror galvanometer on a stone shelf was the main antagonist.

Demonstrators offered two main reasons for such changes. First came personal interest: an idea or effect which had caught their fancy. Second came things in which they had found students weak, to reinforce knowledge of them.

However, in devising the new exercise, the main problem may lie not in the physics, but in contriving something which will offer most students a useful experience which is not necessarily the same from student to student. For it is often the feeling that he has achieved something useful that keeps a student studying. To bring that off, the demonstrator may need to project himself into the minds of others in a way that needs remarkable qualities that few feel sure of possessing. Not many originators of exercises would claim that they very consciously used such insight, but most experienced demonstrators very quickly sniff out an experiment that, clever as it may seem, is not a goer.

So it could happen that a new staff demonstrator, keen to contribute to the laboratory some exciting and valuable new experiment, produces it but has to be told that it will not do because it is too long or too short, too difficult for weak students or too easy for good ones, or some other more complicated reason. Responsible physicists have told us of such experiences from their youth which are consistent with such a picture, which from one side is of obstruction and inertia, but from the other of judgement, foresight, and experience.

The purpose of this section is to show, not reasons for antagonism, of which we saw little, but how the structure of a large teaching laboratory has a stabilising mechanism that keeps it much the same over long periods of time, evolving but rarely suffering revolutionary changes, when it is the latter which might be expected in view of the steady influx of new, clever, critical young men. This stability can be undesirable. Two large departments we visited had introduced or evolved systems of subdivision (called 'unit laboratories' at Birmingham - see 6.4 and 8.4; see also 5.6), which made change a more reasonable expectation.

More generally still, the difficulties exposed in the whole discussion of staff demonstrators and their roles explain why the discussion has a delicately ambiguous tone. Little in the actions of demonstrators can be truthfully delineated in black and white.

2.4   THE POST-GRADUATE DEMONSTRATOR

In the much despised monitorial system once used in schools, economy of teachers was achieved by having advanced pupils pay for their continuing education by helping in the teaching of younger ones. The use of post-graduate demonstrators in laboratories has some remote analogy, at least in that it is not easy even to decide whether to look at it as a kindly device to help them through lean years by modest payments, or as the exploitation of cheap labour to avoid the need for more staff or more calls on staff time.

If, however, there are tensions of this kind, they are kept well below the surface. Staff do from time to time express doubts about how effectively post-graduates can be expected to teach; doubts often countered by pointing out that they are likely, being much of an age with students, to understand their problems better than staff. In the laboratory, one sees them moving about their appointed tasks with little obvious uncertainty, with perhaps just a suspicion on the part of staff that if one of them disappears for a time the reasons are best not inquired into. Whether the post-graduate demonstrators are good or not is often treated as a matter of luck:

> 'They are above average in this laboratory, because they put in a bit of work, and they have some knowledge.'

Normally, it is assumed that the demonstrator will learn the job by doing it; training of any kind is not so much considered and rejected as not considered at all, on both sides. So a post-graduate will say:

> 'Demonstrating can't be taught - you just have to pick it up.'

This one felt, for instance, that there would be no point in discussing marking of reports with examples, even though he had considerable doubts about his own marking. Thus the view taken of demonstrating is that often taken of university teaching in general; that it is not discussable except in the most general terms, and that a man is to be expected to bring to it his own special qualities, tempering them to fit tolerably into the existing system.

Post-graduate students tend to agree with the general opinion that their special positive contribution is closeness to the student:

> 'At least we remember how they feel.'

Several made equivalent remarks, though the one who took special trouble to take his group of students off to his research laboratory to interest them in his work and break the ice was a heartening exception to the rule.

Students tend, but not so decisively, to see post-graduate demonstrators as less threatening than staff:

> 'They are much the same really. They ask the same difficult questions, but I think you feel you can be a bit

more flippant with the research students.'

'I prefer a research student. If I go to a staff member
it will be a teacher-student relationship and not a free
discussion - you can't talk as much. There's a barrier
there.'

In terms of value for money, students do not invariably let
comfort distort their view of who best delivers the necessary
goods:

'The research students don't know the problems anyway.
You can't really blame them; they have so many
different experiments to look after.'

'The doctors*... tend to be after the right things, in so
far as they're after your understanding of the experiment
... whereas the postgrads... might tend to concentrate
on your error calculations.'

One common justification for using post-graduate demonstrators is that they can give value by thinking on their feet,
demonstrating as they do so the approach to experimenting
which, as research students, they are currently experiencing.
One gave as an example:

'As a post-graduate I do research, and I know simple
things like techniques for keeping records, which I can
teach them.'

When it comes to direct help, sometimes students find staff
better and sometimes not:

'The postgrads generally don't know a thing about the
experiment and have to start racking their brains from
first principles, and it could take ages... We found
we could do much better... by waiting for someone who
knows about it.'

'Sometimes the post-graduate demonstrator is very good.
He's done the experiment before. The staff are slow,
sometimes, in working out problems.'

Both students may be wrong. 'Racking their brains from first
principles' could be the very thing most worth doing, and

----

* 'doctors', staff may like to know, is a widely used term of
differentiation.

remembering how to do the experiment from two or three years back the worst.

The last comment points to an assumption of some importance: that the post-graduate demonstrator has been through the laboratory himself. More than one laboratory director spoke of this assumption, pointing out the problem of research students who came from other universities not knowing the laboratory. On the surface, the problem is their not knowing the experiments, but behind that can sometimes be sensed that the problem is as much not knowing what the experiments are for. This issue is rarely clear, however, since it is taken for granted that the purpose of the laboratory is ordinarily something to be absorbed, not something that can be explained in any detail.

In a laboratory which has slowly evolved over time, this implicit nature of its purposes must be in some degree inevitable. All sorts of little adjustments and arrangements have been arrived at by making it run more smoothly in a variety of respects, on grounds more often pragmatic than theoretical. It is in just these laboratories that one finds the exact role of post-graduate demonstrators at once taken for granted as fairly obvious and regarded as something to be picked up on the job. It was in those laboratories which have undergone some substantial change, or in those newly set up in new universities, where we found their role being questioned.

Sometimes, this questioning goes as far as exclusion. Some staff responsible for newly restructured laboratories take the view that post-graduates cannot cope with the teaching required. However, looking at examples of laboratories which began newly five to ten years ago, one does find post-graduates taking in them much the same role as they do in those with a longer continuous history. The reason is the same: having been through it, they are supposed to know what to do.

Notably, the one example we found of systematic training of post-graduate students was at Sussex (see chapter 5) where the course served by the laboratory called (because of student numbers) for the immediate availability of some forty demonstrators, none of whom could have passed through the previously non-existent Sussex system.

In long-established laboratories, although it is said (not without pride) that matters are organised slightly other than they are elsewhere, in general the job of the post-graduate demonstrator is taken to be much the same as at any other university, details of experiments apart. In fact, however, there are considerable differences.

In some places, staff reserve for themselves jobs like marking reports, leaving research students to assist mainly with matters of nuts and bolts. In others, staff do little or no marking, and exercise only a general supervisory role. In not a few, both staff and post-graduates share equally in all tasks, it often being left to the student to decide whom to approach both for help and to have his work marked. Teaching in the laboratory varies too. In some places staff teach actively while research students stand by to fill in gaps; in others it is the other way round. In yet others, both are always busy, or both let well alone.

If only it were common to find post-graduates confident in their work, and staff fully satisfied as to their effectiveness, such differences and the lack of much attempt to guide or help would not matter. In fact, though, post-graduate demonstrators quite often are unsure in their touch. As they say:

'It is difficult to get them to enjoy the experiments.'

'Sometimes I try to explain the theory to them, but they don't care at all.'

'You're slogging away trying to get anything at all out of them.'

This is not to say that they do noticeably badly. The impression we have of most laboratories is that staff and post-graduates are much more on a par as far as students are concerned, than that they differ. Comments on the differences such as those quoted previously are normally qualified, and the remarks above might easily have come from staff. Many post-graduate demonstrators visibly do a good job, even when they feel unsure about it.

What could perhaps be useful is for staff and post-graduate demonstrators to discuss matters rather more, despite the hesitancy of both as to whether laboratory teaching is discussable. Such discussion could at least convey relevant information, so that it would be less common than it is to find a demonstrator saying that he did not know if errors were taught in the previous year, not being a graduate of that university. Marks and standards could be talked about a little, with examples, which might ultimately help students to feel that they knew why they had the mark they did. Those post-graduate students with a real feeling for students, and for the kind of help they need, might be able to communicate something of it to others, including staff.

23

2.5   STUDENTS

Students, besides outnumbering all other laboratory inhabitants, differ widely amongst themselves, and in ways that may take some time to show.   It is perhaps in the laboratory, though, and certainly more than in the lecture room, that such differences can most easily emerge.

One thing that many share, though in varying degrees, is some initial anxiety when they arrive in the first year.   It shows up, for example, in what one said when asked what he would say to next year's students about the laboratory:

> 'I'd tell them it's not going to be as bad as they thought it would be - that they can more or less take it in their stride...'

and explaining, mentioned,

> 'Everything you hear about the university, and first year students there - that they're thrown straight in to complicated experiments, and get a bit lost.'

'Can I stand the new pace?' is what all must hope will prove to be the case, and what few can be certain of.

To begin with, it is often other students who are suspected of being altogether more advanced, clever, or mature:

> 'I felt as if I'd be at a disadvantage right away, and I thought I'd hold everyone back by questioning.'

Such feelings fade.   The oscilloscope which, in November made one feel a fool by not yielding a trace, was, by the Spring a normal thing, and by the second year, its behaviour might be evidence of meddling by some third party.

It is notable how second year students assume as obvious their gain in confidence, and regularly assert that what is done in the laboratory in the second year could not have been done in the first.   They say this regardless of what these things are, and are not impressed by any counterproposal to the effect that in other places what they do in the second year is in fact done in the first.

While the change owes something to growing familiarity, it probably owes as much to talking with other students.   When they find that bafflement is widespread, and that its incidence is more nearly random than systematic, it comes to seem less of a personal threat and more part of a way of life, one even to be treated with irony:

'Don't those mu-mesons strike terror to your very soul?'

Along with confidence comes a dislike of routine and a growing preference for being left alone to work things out:

'I think it's better to leave the experiments with the difficulties there. I think you probably learn more. It's no use just going up to the equipment and being told to take this measurement. Anyone can do that.'

The shade of arrogance detectable here tends to appear when such students have successfully confronted and defeated a number of such difficulties. New tasks, however, can be almost, if not quite, as daunting as before:

'If I know exactly what I'm doing I shall be OK, but if I have any doubts I shall be very wary of starting anything.'

Routine is often disliked for the good reason that it implies an absence of taking thought or of initiative,

'The boredom of going into the lab and... getting results but not achieving anything in yourself, because you've just followed someone else's instructions... But if you've got to think for yourself you become so much more involved and it's better for you.'

but students were capable of switching into the other mode, of wanting to be told what to do, if the demand seemed too great:

'I had no idea what I was doing, and the manual didn't explain what you had to do - very waffly, tells you a lot of theory, but coming down to apparatus it didn't tell you anything at all.'

A sufficient number of students, both first and second year, were bistable in this respect for it to appear that finding out which state a student is in at any one time, and reacting accordingly, is one of the more difficult tasks facing a demonstrator. Triggering a changeover can be harder still.

From discussions with students, we gained the impression (though it may owe something to unfounded prior expectations) that it was the women students who felt most acutely the fear of helplessness at the start. At least they it was who

25

frequently expressed it in stronger terms, and who gave good practical reasons for it such as a previous relative lack of acquaintance with apparatus. Male morale seems to be a little more impervious to unobserved failure, though it can be questioned whether this is a matter for congratulation.

2.5.1 THE 'GOOD' AND THE 'WEAK'

That some students are better than others is not a proposition that many would question. 'Good' and 'weak', which are widely used terms, might however be thought to be rather too bipolar to describe well what must be a continuous distribution.

One reason for accepting these terms is that it is only necessary for a visitor to talk for ten minutes or so with a student to form a clear idea of whether that student would class himself as amongst the better or the worse. Perhaps the strongest reason for supposing oneself to be 'weak' is finding that one often does not see the point of things, or understand what is going on, while others do. By contrast, the 'good' student is relatively unselfconscious about all this, assuming that he knows what he is doing, that its effect will be beneficial, and that things as they are are pretty much as they ought to be.

If this account is correct, so that it is more often the 'weak' who complain that instructions are inadequate, that the taking of readings is endless, or that the theory is incomprehensible, a reason for the difference polarising itself becomes clearer. The staff were, of course, amongst the 'good' ones, and they naturally identify with and are identified with by, a new cohort who share similar views. The others stay out in the cold.

It is striking how, in going from department to department, the proportion of students described as 'good' to those described as 'weak', does not vary very markedly, despite considerable differences in the quality of students they say they attract. That is to say, 'goodness' and 'weakness' owe at least something to the way attitudes work on students, as well as to their actual qualities of intellect or drive.

One possible mechanism is in the response of staff to changes in how the laboratory seems to work. Thirty years ago, students used to measure the change in pressure of a gas with temperature; now we saw no experiment like this, and it is changes in students which have brought the change about. One might guess that if slowly students appeared to be finding the same experiments easier and easier, they would be altered until the expected proportion of 'good' to 'weak' students was approximated again. The remark is at most a reminder, not pure cynicism. In so far as the university sets intellectual standards, such a mechanism is needed to get the best out of those who come, while ensuring both that the flow of the best does not dry up to a trickle and that few are utterly defeated by the demands made on them.

To a considerable extent, it is the 'good' students who can and do play the system. They find out how the laboratory is organised. They discover which experiments might interest them, and which might give most credit for reasonable effort, so ending up doing work which pays off doubly, in having been done well because of the attention and effort brought by interest. Such calculations are merely an extension of physics, rational behaviour it would be negligent to omit. Indeed, the laboratory has actually been organised to provide exactly these benefits, not to baffle or annoy. Its advantages are meant to be taken, and it is the 'good' students who take them, besides not infrequently being very capable and working hard.

The successes of the system in its own terms justify it. If there is a danger, it is that it takes too little account of those who are misfits in it. At least some 'weak' students (like Einstein) do not want what they are supposed to, and do sometimes want what they are not supposed to.

2.5.2 THE ACCEPTANCE WORLD

In almost every case, far from rebelling or reacting strongly against laboratory work, students implicitly accepted it. That is, while critical of detail, they tended to see it as a job to be got on with which could not readily be other than it was. The tone comes through, for example, in:

> 'You've got a certain number to do. We've been spending about two weeks on each, but we've been getting bad marks, so we decided to spend the proper time on this one...'

with the student then going on to say,

> '... this experiment here, you could spend days on it looking at all sorts of aspects. You decide on the length of time it will take, and cut down what you do to work within that.'

Acceptance is perhaps more marked, and less tinged with opportunism than is the above, in the first year:

> 'It was a good idea to have a gentle start, though I thought at first it was too much like school. Now I see that if I had had to do the later experiments straight off at the start, I would have been demoralised.'

What strikes the observer, going from one laboratory to another, is the very considerable differences between what students accept as natural and inevitable. So, like the previous one, one may say·

> 'I think it helps to bring everyone to the same level. Some seemed rather easy - I had used the oscilloscope before, but others won't have.'

while others will take it as quite proper that the first year should be a radical break. In one place it might be a new and tough regime of measuring physical constants, and in another a series of more or less open project-type experiments. In one like the last, a student could say:

> 'It's much freer. You devise your own experiment. You know roughly what should happen, but you're on your own.'

It is not surprising that students judge (or rationalise) what they find very much on its own terms: it is what they know. But it does mean that staff in two very differently organised laboratories are each very likely to claim, with truth, that students seem broadly satisfied with the way things are.

2.6   A TECHNICIAN OR TWO

On sporadic occasions, we came across the tip of the iceberg of the influence of the technical staff of laboratories. It is pervasive rather than clear, but it still seems worth drawing attention by example to the possible importance of the technical staff.

One of their most evident importances is in the quality of apparatus. In one place, where the laboratory technician belonged to a departmental apparatus construction group, which had the support of the laboratory director, the standard of the apparatus was the thing which first struck the visitor, both in its design and finish, and more importantly, in being well contrived for its specific teaching job.

Students not infrequently use the technical staff as demonstrators:

> 'He puts you right. For example I had a switch on the bench I didn't know what to do with. I would have spent half the afternoon looking for how to use it.'

The technical staff often have a permanence in the laboratory that cannot be matched, coming to regard it as their own, and looking tolerantly on the transient attempts of staff to introduce changes. Thus one confident young man said:

> 'I know it inside out. I enjoy showing students how things work, having a joke with them, and getting them thinking.'

He showed an admirable grasp of the physics of several experiments, and had clear views about the point of the experiments:

> 'They are designed so that if you do it you get some sort of mark. But a good student can get a hell of a lot more out of it -really go into it. For example, that B-H loop: a good one might get a loop and wonder what would happen if you made the specimen smaller, or changed the geometry. A poor one gets the loop and packs it in.'

It may be worth noting that in this laboratory, staff, postgraduate demonstrators and the technician met from time to time to discuss how satisfactory the laboratory was.

Technicians are not under any illusions about staff:

> 'You've got good demonstrators and indifferent demonstrators. Some take their work more seriously than others.'

One described qualities he had found and valued in laboratory directors:

> '... a mature man, a man with patience, a man with a lot of knowledge which the students can learn from, rather than a more impetuous but perhaps more with-it younger scientist.'

He had seen the laboratory change appreciably, but took a solidly reasonable view of the changes:

'... when the change came I was getting a bit old, so I
resented it a bit, but it was a good idea and I think...
it benefited the students and benefited me in the end.'

Such a man can exert a powerful influence, in a teaching role
too, if he wishes:

'I get a bit involved, probably, over and above the call
of duty... I keep some good write-ups, and sometimes
if there's a pair who can't express themselves, I say,
"Well, look, so and so did this, and this is a good
write-up. Don't copy it, whatever you do, but it will
show you the kind of standard you're expected to reach."
Also sometimes it's good for the demonstrators... The
imported postgrads, it's hard on them. They come here
not having a clue what the experiments are all about.
So if I can produce a couple of good write-ups, that
gives them a start. It helps me too - for instance, our
X-ray experiment has gone a bit kaput... so I found
some very nice photographs which were taken, and when
I put in a new crystal and start setting up I shall know
what to look for.'

Such men are no doubt rare, but it is good to know that they
exist.

3. Processes

This chapter concerns things which happen in laboratories. They are of many kinds, but all are linked to what students learn, and to how they learn it.

The discussion is based on, and quotations are taken from, material noted or recorded in the course of visits to a number of teaching laboratories. For the purpose of this chapter, the identities of the laboratories are unimportant, and they are therefore disguised. In general the discussion is restricted to things which can happen in laboratories organised in one or other of a few common ways, and excludes things which are likely to be associated with special or very unusual arrangements.

The various sections which follow each look at events from a particular point of view. These points of view could have been chosen differently, but taken together do seem to us to represent something of the variety of ways in which one can think about the processes involved in laboratory work.

3.1 DOING A LONG EXPERIMENT

In many laboratories experiments get longer as the under-graduate course progresses: perhaps occupying laboratory time for a week in the first year, three or four weeks in the second, and extending often to a project lasting a term or more in the third.

The intention of this shift is often explained in terms of realism, and - related to it - of providing time for students to make their own decisions and carry them through. Thus, correspondingly, the instructions for experiments may - but do not always - get shorter or less detailed and prescriptive as the experiments get longer.

In one university the shift from first to second year was intended by staff, and was certainly noticed by students:

S 1   In the first year, they say do this - take this measurement -

S 2   Although they say they don't

S 1   - divide this measurement by two... and write down your answer.

S 2   They do say in the scripts that that is not what they do ... if you want to make some modification you can, but you must consult them first.

S 1   You can't do it in five hours anyway - it's impossible - so it takes a lot of the interest out of it. You can't say, well this is a better way of doing it; you've got to do it their way.

Indeed, in the second year laboratory, time was often mentioned by students as crucial:

> 'You have time to think and prepare - time to get to see the problems.'

> 'You're not pressurised - you can go on until you've done as much as you can... you can deviate and have time to think.'

Six to eight experiments were expected in the year, with each occupying twenty to thirty hours spread over three to four weeks, with students given some right of decision as to how long they spent on each.

How is the time spent? What do students do, and what of it is of value?

Many begin with rather little idea of what they are in for, choosing an experiment on grounds such as that it was vacant, seemed 'vaguely interesting', or that they had heard from others that it was not excessively tiresome or baffling. Similar reasons were given by students in a number of laboratories, with a preliminary discussion with a member of staff about what to choose or what best to do with the chosen experiment being the exception rather than the rule.

Typically then, a pair of students might spend an hour or more 'to decide what we had to get out of it' (note the nice balance of freedom and compulsion in the phrasing). In this laboratory, as in several but not in all, scripts gave only references for the theory, so that three or four hours might then be spent in the library in laboratory time. A member of staff would sometimes be called in to advise: for example to judge whether they ought at that stage, before having met it in lectures, to grasp all the theory or only its outline and results.

The next phase is most typically described by students as 'trying to get it working'. Indeed a week or so may pass in getting over the initial difficulties.

Perhaps much to be hoped for at this stage would be some trial experiments, and an exploration of relevant conditions, so as to plan the later collecting of data in a way which could minimise errors, both random and systematic.

Inevitably though, in fact such principled activity is mingled with crucial if less profound things, notably making the apparatus work at all - getting photographs which are other than blank, for example. And for the student, the difference between an important source of systematic error which deserves long thought, and a trivial malfunction which is soluble quickly only in outside expertise, is rarely obvious:

> 'We had to completely rethink that one, because we thought we were doing it wrong where in fact the apparatus was all out of line.'

Whether the outcome of such problems is, and is seen as, good or bad must depend a good deal on the acuteness and flexibility of staff:

> 'We had so much go wrong with it, we spent half our time deciding why it had gone wrong and how that had affected our results, which maybe they thought was as good a substitute for the actual result they wanted.'

At all events, these students thought it good:

> 'We had to think an awful lot about that experiment, and really consider exactly how each part was interacting with the next one and the one before.'

Observed from the outside, the 'getting it working' phase can easily look, and may sometimes be, aimless. Students are to be seen trying one thing and then another, being diverted by some small need such as obtaining a screwdriver, talking (often desultorily) about what is or is not going on, and so on. It is, then, the phase which presents the demonstrator with the greatest challenges to his intuition and tact, and one where action or inaction on his part can both be the best or the worst course.

Next, since the overt goal of such experiments is usually the measurement of some quantity, there is a phase of 'getting the results'. Again, this can occupy laboratory time for a week -

up to perhaps ten hours.

It is characteristic of long experiments that few concessions are made to the demands of the physics. If the experiment requires ten runs of repetitive collecting of twenty values per run, so be it, is what appears to be the general attitude. By contrast, shorter experiments cannot afford the same luxury, and their apparatus is not infrequently ingeniously devised so as to eliminate a source of error which would, in a longer version, be left to contaminate the data and require the collection of more to eliminate it.

Not surprisingly, it is here that students speak of boredom, though in a way which acknowledges the soothing nature of repeating a well known act many times:

I   And doing the experiments consists of what?
S I   Taking various lengths of oil... timing for so many
      graduations. That was the boring bit.
S 2   It was - very boring. Sitting there watching the little
      scale pan going down, revolving with different weights.
S 1   Different weights as well, yes...
S 2   ... we did about fifteen different readings, each one
      giving us a value of the viscosity of the oil. And then
      we averaged over those... Took a mean standard
      deviation, which we worked out wrongly... because I
      used N instead of $\sqrt{N}$, which wasn't very clever.

The link with the final phase, that of writing up, is strong. The number of results taken depends, for many students, on what is required for an acceptable analysis and error estimation, always within limits set by the time needed to collect and process data.

Here the claim of 'reality' deserves some thought. A research experiment may demand hundreds of hours of data collection; in a teaching experiment, the requirement is not set by the current state of knowledge, and can be and is adjusted to the weight put upon laboratory work. At the same time, experiments and apparatus are often designed so as to make this requirement seem more or less a natural outcrop of the problem, not merely a pedagogic constraint. It is in areas like these that one hears most double-talk between staff and students: staff explaining that they really ought to have looked into a certain effect, and students parrying with arguments such as that they thought it relatively unimportant, meaning perhaps that they judged it not worth their time in comparison with other things.

34

A considerable time, from perhaps three to twelve hours, may be spent on calculations and writing up. Here the effect of the ready availability of calculators and often of computing facilities, makes itself felt. Quantities of data which would have occupied a student for a week twenty years ago, and so would never have been collected in the first place, are not infrequently taken and processed. The main limitation is the time needed to collect it, which is inclined to depend on the topic of the experiment. No doubt, though, the students who said,

'We had a spare half hour so we came up and took some more readings'

were influenced in their decision by being able to cope with the extra data.

The laboratory from which most of the above material is taken is typical of laboratories containing mainly three to four week experiments, but untypical in that its policy is clearer than in several other places. Thus, while elsewhere, experiments vary widely in how much theory they give, how much advice, and how many detailed suggestions of what to do, as well as in the sophistication of apparatus, in this laboratory many such variations have been removed. In other laboratories, then, the picture of activity one sees is more varied, with the main phases described above overlaid with others produced by the circumstances rather than by the nature of the task set when a student settles down to spend a month measuring, say, the elastic moduli of glass.

A useful metaphor for thinking about such practical work may be 'flying time'. That is, although a prospective airline pilot is given much theoretical training, and also spends many hours learning a host of particular skills, an essential ingredient of his training is simply time spent flying. The reason is that the real activity for which he is being trained can be relied upon, given time, to turn up situations which have a sufficient variety of the unforeseen about them, so that the trainee's use of his skills can become more and more immediate and appropriate. Similarly, one main justification for having students simply spend time in the laboratory trying to do things with apparatus, on whatever pretext, is that those pretexts will inevitably turn up at least a sample of the kind of difficulties, foreseen and unforeseen, that any experimentalist will meet.

Time, however, is limited. The central dilemma posed by such experiments for the design of a teaching laboratory is that they rely upon happy circumstances to produce enough good

things and not too many irrelevancies.   It is not easy to judge
what is a sufficient density of occurrence of good events:

> 'You spend a lot of time and you don't seem to achieve
> a great deal in the time... but then I haven't given up
> and walked out for a fag today - sometimes I have to.'

Some, then, will prefer to design experimental work more
tightly around particular activities (see especially the discussion
of a 'unit laboratory' in chapter 6).   That decision involves
opting for less of leaving students well alone, and for less of
naturalness about some experiments.   But it recognises the
potential self-deceit of those who argue that whatever turns up
in an experiment must be good just because it derives from the
experiment as 'real' physics.   In fact, every experiment is part
natural and part educational contrivance.   The problem is to
contrive them to produce roughly the right number and kind of
difficulties, in circumstances in which students react to the
difficulties as having to do more with doing physics than with
coping with the pedagogic set-up.

3.2   LEARNING ABOUT APPARATUS

On a time scale shorter than that of the previous section,
another aspect of doing an experiment is what is often called
'meeting' equipment or 'getting experience' with apparatus.   The
process is not often more deeply analysed, not least because a
little reflection suggests that it is both complex, and very
variable as between people.

One problem is that it is hard to say just what is the content
of what is learned when one becomes more adept with apparatus.
If the content is thought of as like the list of ideas in a lecture
course, practical work at once seems very slow.   The time
occupied by three lectures - perhaps ten percent of a whole
lecture course - can be used in the first attempt to get, say,
a direct-reading potentiometer connected up and giving results.
As one student said:

> 'I suppose it depends what you think is money for time.
> If you think of spending an hour and a half learning...
> a technique, then I suppose it's worthwhile.   I don't
> think you could learn it any other way - without trying
> it.'

One small but important difficulty is that small but important
things can go wrong:

'She kept getting blank plates. She's still doing her experiment now, even though the labs are finished... First two plates were blank... then they ran out of plates, and now they're using a different developer so that they have to wait an hour and a half for it. Just about everything that could possibly go wrong goes wrong.'

Nor, as noted before, is it easy for the student to tell whether what is wrong is relatively trivial or important, in terms of learning something useful.

To look a little more closely at what might count as success or failure in practical work intended to promote facility in using apparatus, there follow descriptions of the work observed being done by two pairs and by one single student in one laboratory.

The student working alone had an integrating operational amplifier, the means for calibrating it, and a search coil and magnet on which to use it to measure flux. After a quarter of an hour, he had plugged in the amplifier, but had only looked at the rest of the apparatus, comparing it with circuits in the script. After half an hour he had connected one resistor into part of a circuit, and had looked without evident comprehension at the movements made by the pointer of a voltmeter as he turned the two main amplifier controls. At this point the demonstrator (working alone with twelve students) approached unasked, and found by questioning that he did not know how to connect the circuit, nor which part of the experiment to begin on, and spent ten minutes connecting up and getting the circuit and amplifier working. Through this, the student stood silently by.

The nearby pair had a difficult experiment, and had, after an hour and a half, reached the point where it was connected but where they knew that it was not functioning as it should. It happened that the demonstrator was busy dealing with a broken-down instrument elsewhere, and they waited, with several attempts to gain his attention, for half an hour, after which he came and in a minute found and rectified a faulty earth.

A rather more detailed account of the work of the second pair may be of interest, depicting the pattern of their work over three hours. They had an x-y plotter, and a diode, with instructions to obtain the forward and reverse characteristics of the diode (that the reverse impedance was larger than that of the plotter was a deliberately laid trap). In addition, they had a low frequency triangle signal generator, and a power amplifier, and the script gave basic data about all the instruments and some brief suggestions about circuitry.

The first five minutes was spent looking at the script. The next twenty five minutes was occupied in connecting up, trying out the controls, and learning to recognise what was going on. This went with continual relevant comments from one to the other: 'That's the slow variation (of the signal generator)'; 'You've got to connect that to that'; 'There's a very small current here'. They put in practice, making for example several dummy runs with the plotter until they could control it.

After half an hour they connected in the diode, and got a trace with a right-angled bend:

Recognising that it was not what they wanted, they talked about why they had such a trace, doing calculations aloud: '2.5 volts divided by 5 ohms which is $\frac{1}{2}$ amp - that's far too large'. The problem was still unresolved:

S 1   Think, think, think.
S 2   It's the sensitivity.
S 1   I'm not worried about that - I'm worried about the resistance box.

In a quarter of an hour, this led to their getting the expected trace:

After trying it several times they went to get graph paper for a final plot. The demonstrator approached unasked, saw without asking where they were, and asked if they knew where the zeroes (axes) of the trace were. They suggested a correct way to locate them, and he went off, the exchange having lasted three minutes.

Thus, after fifty minutes in all, they were now ready to take a trace. But the trace then developed a loop, the x-axis coming out double. It took seven minutes to discover that this was an earthing problem ('I pulled the wrong rope'). The trace still had a loop in it, though, and they 'solved' this by lifting the pen before the 'error' could occur (That's it, then').

A comparison with a trace obtained by a previous pair showed some differences, and they set to changing the sensitivity of the plotter, again doing calculations aloud ('Ah - I see why - I'm not thinking'; 'We added on a forward current - we must have done'; 'I think it's too sensitive'; 'What's the breakdown voltage?'; 'It's affecting the zeroes, which I can't understand').

This occupied twenty minutes, during which the demonstrator passed by, looking at what they were doing. They ignored him, and went on arguing.

Now they decided to try the diode in reverse, and spent another quarter of an hour trying to get that working, again arguing about the traces. One walked off without comment, and came back having found from those who had done it last that the input impedance of the plotter was much the same as the reverse impedance of the diode.

After altogether an hour and a half, they decided to go back to the forward bias, and get proper traces, taking half an hour to do so. Then, in the final hour of the three, they got the reverse bias characteristic, again with calculation and thought:

S 1   5 volts across here - that's about what we want.
S 2   What's the voltage across this resistance?
S 1   The same.
S 2   No, all the voltage is across this one.
S 1   Switch off and have another think.

By the end of three hours they had both graphs, and seemed to understand them.

It seems reasonable to argue that much of what went on in this last instance was of value. In such a judgement, small details are important: the insistance on taking thought; the initial practising; the meeting and overcoming of a variety of problems, each within a modest span of time.

It would seem, that for this pair, the exercise set was well judged. It occupied about the right time, and posed problems within their reach. The script appears to have given just about the right amount of information, leaving the students to do a well-defined but not purely mechanical task.

None of this, however, guarantees success. The pair described before appeared about equally competent, but hit a minor but crucial snag which they could not deal with, at a moment when the help they knew they needed could not be obtained. In the context of ordinary life, half an hour is hardly a long time in which to learn by experience that earthing electrical instruments can mean the difference between total success and failure; in the context of a planned laboratory exercise it can look excessive, whether it is or not.

The first student described, who worked alone, shows how hard it is to judge exercises and instructions which will suit all students. Where others spent half hour blocks of time on parts of the problems set them, he spent one getting essentially nowhere. The difficulty is to decide whether the lessons learned about the need to know much more about the instrument were the best use of that time, or whether the event was inevitable with this student. But he did get help, and within not too great a time. Possibly he could have learned more from the kind of help he got, but it at least got the apparatus functioning. Nor could the demonstrator have told in advance whether the malfunction was one of equipment, or, as it turned out to be, of intellect.

One lesson that can be drawn is that close planning, whether of the exercise set, for help to be available, or of routines for deciding when help is needed, cannot avert or foresee all the inevitable problems. Problems can be trivial or major, but either can bring work to a halt. Some students may appear to need help (as when the last pair described got a looped trace) but decline it and do (as they did) or do not sort it out themselves. Others who get help might, or might not, have managed without.

Another lesson is that such exercises look more valuable when they provoke thought, besides leading to the successful completion of the immediate job. It is too easy to see them as the practice of manual skills, and they can too easily degenerate into following instructions.

The necessary short time scale of such exercises does, however, constrain the kind of thing they can achieve. It is not easy to make them meaningful, and the job of assembling the skill acquired into a more complete performance remains to be done elsewhere.

3.3   DEMONSTRATING

'Demonstrators are encouraged to walk around the labs, to ask what you are doing. Don't cold-shoulder them; they are trying to get to know you; besides, they can often make very helpful suggestions.'

This extract from the notes issued in one laboratory illustrates the delicacy of the demonstrator's job, gaining poignancy from the fact that in that laboratory the demonstrators conspicuously did not walk around the labs.

Just what do demonstrators do, and is it in accord with various views of what they should be doing? The following three laboratories may serve as a useful basis for discussion. The neutral tone of the descriptions implies neither approval nor disapproval.

In laboratory A, for first year students, one post-graduate demonstrator finishes showing a pair how to connect a circuit, and goes to the central demonstrators' table and resumes reading a novel. Five minutes later, a second post-graduate fetches a member of staff over to an experiment which is not working, and he obtains the required hysteresis loop, showing how its shape can vary, while the two students and post-graduate look on. Ten minutes later, the first post-graduate gets up and looks over the shoulder of a student working near the central table; the student asks him about the value of factorial n when n is zero. The post-graduate then walks round the lab for a minute, but neither accosts nor is accosted by any student, and so returns to the table, joining the second. After five minutes, one goes off to the technician, and the other approaches a pair of students unasked, and spends ten minutes trying to make sure they understand the main point of their experiment. These activities are what occupy the two demonstrators for half an hour during the afternoon, in a laboratory containing some twenty students.

In laboratory B, for second year students, one of the two post-graduate demonstrators sharing responsibility for a group of about ten students in one part of a larger laboratory, gets up for a walk around. He approaches one pair:

D   Have you started taking measurements yet?
S   Yes.
D   Of what?

The students explain, and are asked a factual question. When one student looks in his notebook for the answer, the demonstrator says it does not matter, and asks instead how they are locating nodes in a standing wave:

D   Are you measuring once off, or measuring either side and extrapolating?

One student claims that that is really what they are doing, even though it seems as if they are not; the demonstrator presses on them the virtues of a more explicit method, saying that they are throwing data away. He asks some more questions, one about the physical reason for an effect and another about how they had decided on a method, and walks away.

Approaching the next door pair, he asks what they are measuring, but is told that they are not sure they have the right circuit. In fact their oscilloscope is displaying mains hum, and they have not been able to understand why the frequency calculated from the trace is so low. The demonstrator diagnoses earthing troubles, and himself rebuilds the circuit, using the diagram in the script, doing this in silence except when one student points out an error.

Having finished, he goes round the corner and approaches a third pair, who are collecting and recording data. He suggests that they should be plotting a graph as they go along:

D   There are two of you. You've got to think what the results mean... you put weights on and off, but you can't tell without a graph if there's a systematic shift. You can also start asking yourselves which values you believe.

Having made some further criticisms of a technical kind, he leaves them to go back to the demonstrators' table, saying,

D   I think I'll go and have a sit down now.

In laboratory C, again second year, there is a staff demonstrator alone with twelve students, doing four different exercises, some singly and some in pairs. He is constantly on the move, or standing looking around. A few simple categories serve to describe how interactions with students begin:

request for help

approaches unasked
   - diagnosing trouble
   - routine check

It is also worth distinguishing those which begin with students nearby, from those which involve moving to students at a distance. In these terms, the demonstrator in laboratory B is always approaching unasked nearby students for a routine check; giving a chain of events well described as 'doing the rounds'.

Over about an hour, the staff demonstrator in laboratory C works with ten students or pairs, so that in that time he sees all those in the laboratory at least once. The ten events, in sequence, go as follows:

approaches nearby student unasked, diagnosing trouble.

request for help from distant student

request for help from nearby student

approaches unasked nearby student for routine check

request for help from nearby student

request for help from distant student

request for help from distant student

approaches nearby student unasked for routine check

approaches nearby student unasked for routine check

request for help from distant student

In more concrete detail, he first sees a student turning a knob and looking at a meter which remains obstinately at rest. He finds that the student does not understand the circuit, and helps him wire it up. He ends by giving advice on what part of the experiment to start on.

A student now comes across the lab to him for a switch, cutting into the previous conversation. Then a nearby student asks how to get the frequency of an oscilloscope trace. The problem turns out to be an incorrect circuit. Unasked, the demonstrator adds some tuition on the use of the oscilloscope.

Nearby is another student on the same experiment. He gets the same oscilloscope tuition, and a check that the same fault as in the last case is not present. Actually it is, and it is rectified.

Now a student nearby requests attention, asking what the switch is provided for. The demonstrator sees that he has not got anything working, and goes systematically through setting it up. A technical fault appears, and the demonstrator tries to get the student to think what it might be, but then tells him:

D   The potentiometer terminal is giving trouble. It's a fault - it's not there for you to 'discover' so far as I know.

Then a student approaches from across the room and is asked how he is getting on. His experiment is the same as the last, and he is taken through several of the same questions about it. As they talk, a further student queues up to see the

43

demonstrator, and says that he can't get a trace on his oscilloscope. The demonstrator finds the fault, and then repeats unasked the checks he had made before on that experiment with
others.

He then turns to a nearby pair, and converses with them
about their experiment, getting confident and expert replies.
Turning again to another nearby pair he asks:

D   How are you doing?   I haven't seen you yet.

They turn out to have a major apparatus fault, and it takes half
an hour to establish that it is a technical repair job. This
occupies all his attention, involving discussion with the students
involved and acting on several of their suggestions, but so much
so that another request for help goes unnoticed for all that time.

COMMENTS ON THE OBSERVATIONS

The examples illustrate a variety of styles and activities, but
also the important uniformity that demonstrating consists of
small events, lasting only a few minutes at a time.

In university A, much is non-event: the demonstrators are
available and signal the fact, but do not necessarily look for
trouble. In the second, the events are brief bursts of teaching:
the demonstrator approaches not with, 'How are you doing?',
but with a question requiring a principled reply, and goes on
probing a variety of points about apparatus and method. Here
much depends on whether his questions are appropriate to the
moment; on whether he can find a useful teaching point in what
happens at that moment to be going on. In university C, there
is more continual interaction, and more varieties of it than in
the others. More students ask for help, and there are more
cases of the demonstrator spotting trouble, a fact probably
connected with the circumstance that in C students were doing
exercises of a definite kind, while in A and B they were doing
more extended experiments.

A related point about C is that there the member of staff
had sole responsibility for the laboratory. He was not just
doing his stint, he was organising and observing the work there.
A very similar pattern is observable in other laboratories where
staff have a similar position; they know more about the experiments, can diagnose faults as they walk by, and have definite
issues in mind that ought to be brought out (note the repetition
of teaching points in the case of C).

Much, however, of what goes on owes a lot to accident and

to proximity. A demonstrator very often approaches students who happen to be near him, and is not infrequently asked for help by students who are nearby, too. Perhaps the students are acting on the advice given by one:

> 'They tend to find out everything. It's best to ask first before he's got a chance to embarrass you.'

It is noticeable that demonstrators do not act as if there just to give practical aid. They take opportunities to probe understanding, to teach points they think important, and, by no means least, to indicate what might be called the ideology of the laboratory. Each example contains at least one instance of the demonstrator trying to get across how students ought to approach the experiment.

It also appears, however, that very often the demonstrator asks whatever he can think of, without premeditation. This could lead to a useful balance of points made, but it could also owe too much to the accident of the moment. Here the demonstrator as a physicist reacting 'on his feet' is in competition with the demonstrator as a teacher with something to teach.

SOME VIEWS OF DEMONSTRATORS AND STUDENTS

Clearly, demonstrating is a balancing act between doing too much and too little. One young post-graduate demonstrator showed considerable awareness of the problems·

> 'It's better if you go to the student, though you have to be careful not to overdo it. I know what it's like to be an undergraduate, but they don't know what it's like to be a post-graduate - they don't know what role I'm playing.'

He felt that it was experience, not knowledge, that was lacking in students:

> 'They lack experience - they direct their energies the wrong way.'

In the same laboratory, another post-graduate took a less complex view:

> 'Most of their problems are rather simple - getting a circuit connected, for example.'

Thus he saw his job as being available to answer questions, most of which could be expected to arise from inexpertness or failure of apparatus.

Some students like to be left alone:

> 'I've never been approached - I always ask. They sort out your problems; that's what they're here for.'

> 'When I get a problem, I go to them. They're not constantly walking around. I like to get on with it myself.'

Some of these reactions have a defensive air about them:

> 'Having them come round and being told, this is interesting, annoys you if you're not interested in practical work.'

but others see things more as the demonstrator might:

> 'He'll just come and watch, and if you're doing something wrong he'll put you right.'

However, not everyone likes to be detected doing something wrong, and there is some tendency to hope that things may come right without intervention, and to avoid too much contact with demonstrators:

> 'You can fool them into thinking everything is all right ... it depends on whether you have had the problem for long or not.'

It was noticeable that in the examples above, demonstrators made much of the running in working with students, raising new issues, asking questions, and so on. Some of this may be due to students finding it hard to cope with the situation:

> 'I tended to sit there and go, "Yes, yes, yes" and half the time I was thinking, "What on earth is he talking about?" and instead of pushing him, and getting it out of him properly, I just didn't deal with it at all'.

> 'I find Dr X very difficult to understand when he gets going - "Well, there's this current going down here, and you should know how much is here, and this resistor here is obviously doing..." It would take me quite a while sitting down and working it out before I knew what was going on... They know so much about it... They say, "Obviously there's a voltage across there..." as if you can see it, as if there's a little

voltage there saying, "Look, I'm...". '

It is worth noting that it was the same man of whom another said,

'I'm glad I had Dr X. He made sure you understood it'.

Another factor mentioned is that one does not like to ask, because of a feeling that the demonstrator has explained it so many times. Indeed, feeling a fool, a feeling the more understandable when confronted by obdurate hardware, is a constant theme, whether congratulating staff on not making one feel a fool or sadly describing the feeling.

Some students appear to want the impossible, though it is hard to blame them:

'It all depends if the demonstrator can see what is going wrong. It's better when they watch for a crisis and come and help then. If they walk up and say, "OK?", you feel you have to say you understand, even if you don't, which is a waste of time.'

Many, however, are more robust and reasonable:

'If you can get over the feeling that you are going to look ruddy stupid, you can go and ask them. I don't think any of them would jump down your throat. They might give you a funny look, but...'

HELP FROM OTHER STUDENTS

The paid demonstrators are not the only people who demonstrate in the laboratory. Students help each other a good deal. One demonstrator clearly did not know whether to approve or not:

'You can't do anything about it. Physicists can ask other people, so I suppose there's no reason why a student shouldn't - though he'll pick up some wrong ideas. But I think the good chap goes along by himself - he realises it isn't worth asking.'

The same unsureness as to when sensible mutual aid shades into illicit 'cheating' is shown, with some more positive comments too, by these two students:

S 1   I think we did most of that one ourselves.
S 2   Yes. This one, we're asking people who've done it before.

S 1　We asked a demonstrator... because we weren't sure at all... she didn't really know much about it, so we tend to ask other people.

I　Do you think this is a good idea, for students to chat to one another?
S 1　Yes, in a way, to swap ideas.
S 2　As long as you don't rely too heavily on it...
I　Have others asked you?
S 1　Yes... we've had some asking us how we interpret our results, because they were getting very dubious photo-graphs...
S 2　It helps really, because you have to keep remembering what you've done...
S 1　And because they asked us points on that experiment that we hadn't ever thought of... it made us stop and think.

Others in the same laboratory were aware that it can go too far:

'In theory you shouldn't ask for help, but in practice you just go around and copy.'

'Some people copy to get better results - it seems un-fair, but they will not do so well in the end.'

Certainly, without raising questions of copying, observations do suggest the fairly frequent use made by students of advice from one another, both about immediate practical matters and about the nature of various experiments.　One may have reservations, but equally it would be a strange thing to try to prevent the growth of such a community of ideas and experience in a university setting.

SOME GENERAL PROBLEMS

As is documented above, much of the time the demonstrator's job consists of doing nothing very much.　As one student put it:

'They did wander past and say, are you OK? and go away.'

It would be wrong to suppose, though, that in doing nothing the demonstrator is necessarily not teaching; the best kind of teach-ing in the laboratory may sometimes be masterly inactivity. Apart from the virtues of giving students the experience, and pleasure, of sorting things out alone, active demonstrating may sometimes be more interfering than constructive, however well-

48

meant:

> 'If they come round and ask you what you are doing every five minutes, and you have to go into detailed descriptions, you can't really get on with the experiment.'

Good demonstrating is not, however, merely doing nothing. As well as striking a balance, which may differ from student to student, between encouraging independence and making sure the student is not quite lost, it has also to balance being accessible and not breathing down people's necks, knowing the experiments well and yet not spoiling their exploratory aspects, letting students plan for themselves and yet making sure they see what is involved, and so on.

Respect for staff, visibly earned, is likely to be an important ingredient. Students in another (first year) lab had learned after one term that one member of staff could be relied on to help with apparatus, while another could not (observation suggested that they were not totally unjust).

The reality of demonstrating is in some conflict with the dream; the dream that it is in the laboratory that the man with research skills is in direct contact with students, and can teach by example what competent experimentation is all about. Observation shows something much more modest: brief encounters of a particular and occasional kind.

Whether these encounters can do more than give immediate practical help is likely to depend on the relationships in the laboratory. The comments of several students above show how tentatively they approach the matter of revealing their inadequacies, and how the demonstrator's status as expert can get in the way of helpful discussion. One laboratory seemed to have established at least one kind of useful personal relations, through the briskness and forthrightness of the staff. The quality is hard to convey, though perhaps some examples may help. Thus when a demonstrator saw a student with a mess of wire, he advised him to tidy it up, and was heard saying,

> '... or a hand will catch it and break something. That's the voice of experience talking.'

Again, helping a student who then complained that he had not dealt with all the trouble, he said, in tones of cheerful challenge,

> 'It's your experiment, my son!'

He obviously enjoyed demonstrating, and complained that it got boring as time went on and students got to know too much about what they were doing. In this laboratory, little inhibition was noticed about asking for advice or help, and little about giving it. Attitudes are catching.

Finally, though, it must be said that the structure of most laboratories is not designed to promote very much by way of long term dialogue between students and demonstrators. Much more, they are organised around long term private tasks for students. The question is, how much of value could be got from changing the structure, without losing too much of the values of letting students get on with definite, but difficult and complex jobs in their own way.

3.4   FOLLOWING INSTRUCTIONS

'The scripts are a potted history of the laboratory.'

a staff demonstrator

It may be a helpful exaggeration to suggest that often the scripts express the policy of the laboratory, and that after that the demonstrators are puppets on strings in the hands of the man who wrote them, who may long since have been translated to higher things. At all events, they do have an important influence, perhaps mitigated (or confused) by having often been written by several hands, each doing its best to promote one amongst several conflicting policies.

The written instructions given in laboratory scripts vary widely in purpose, style, and length, and may vary a good deal within one laboratory. There are laboratories in which there are no scripts, but there are others in which students can be seen still reading some hours after starting work, without having yet touched apparatus. Some scripts give only bare titles of experiments which can be done; others are more like annotated bibliographies; yet others specify step by step what must be done. Some are miniature theoretical treatises; others are more like gardening hints. Some permit, others command. Some hint, others tell.

Any experimental task involves a spectrum of activities. One extreme is to give the student the minimum concerning all of them, perhaps giving only the title of experiment or project. In various ways and at various levels, a script can define work and responsibilities, or give assistance:

by telling the student what to do (relieving him of decisions) by telling the student what he should know (relieving him of a need for information)

by telling the student about criteria (relieving him of deciding aims).

These things can be done, in varying mixtures, within the script or face to face. So the question of scripts here interacts with the whole demonstrating policy of the laboratory, with the range running from places where everything is done in discussion, through ones where much is in the script, but staff work closely with students and tell them a good deal, to those where all but small technical accidents are supposed to have been anticipated in writing.

RELIEVING DECISIONS

One reason for telling students what to do is to help them get started, so that there is soon something productive to discuss. A post-graduate demonstrator saw this initiating function as important:

> 'In most experiments you are given an initial push with the script.'

Then, having begun, he saw it as the demonstrator's job to help get value out of the experiment.

Such reasoning can be carried further, as these students do:

> 'It's very much pre-planned... You wouldn't get any-where if you planned your own.'

> 'Everything is quite well mapped out... That's good because otherwise the standard would have to be lowered. This way we can use sophisticated things.'

While these first year students might speak for many, in wanting a firm framework, and appreciating being able to use complex equipment by being told how to, it is important to remember that first year laboratories do exist in which no instructions, or minimal instructions, are given, and that in them students do get on and work, using apparatus which is not necessarily simple and familiar. Students in one such laboratory sometimes said things like:

> 'It's free... you can let the staff see your imagination ... You're coming up against new problems.'

while some certainly wanted to be told what to do and feared looking foolish:

'I like to get it exactly right, as it should be.'

One reason, then, for not telling students overmuch about what to do and what they might find out is to present the right challenge. A student in a second year laboratory accurately caught the nature of the instructions:

'The basic instruction sheet gives you rough outlines... tells you roughly how to do the experiment, but doesn't tell you the problems. I don't particularly want to know the problems beforehand. It seems simple, but it's a challenge to get the most out of it.'

The staff in this laboratory shared the same view:

'It's up to them to get on with it.'

and this view was reflected in how they acted, and in the document describing the laboratory, as well as in the scripts. The scripts had a consistent style, giving references for all the theory and for much experimental design, restricting themselves to introducing the experiment in a paragraph, and adding up to a page of notes giving hints about matters that needed thought or planning.

It appeared to us that the effectiveness of this policy depended on its consistency. In another laboratory, the scripts varied much more widely in style, and correspondingly students seemed to have a less clear picture of what work there was all about. A typical comment, showing a realisation that the tasks offered many opportunities, but a less clear positiveness of response, was,

'You could spend days on it looking at all sorts of aspects, but you decide on the length of time it will take and cut down what you do to work within that.'

At all events, looking at the scripts in this laboratory, their variety was such that they could easily appear to the student to be heading in many directions. Some were two sides of references and brief notes; others could contain four pages of detailed mathematics followed variously either by a short paragraph of practical suggestions, or by an equal number of pages of detailed instructions. One amusing extreme case consisted of twelve pages of detailed practical instructions prefaced by:

'WARNING: Do not tamper with the apparatus in any
way until you are thoroughly familiar with this script.'

Naturally, students often sense that even though decisions
are left to them, someone somewhere has a good idea of what
they should be:

'I think there's really a set method for these
experiments. You can spend hours on the wrong track.'

giving as an example a hint in the script, 'You will need to
take temperature effects into account'. Indeed, there is an
inevitable tension between the value of letting students decide
and then live with the consequences, and the values one can
foresee coming out of particular things they could be led to do
with particular apparatus. Also, there is the problem of how
far one should go in giving the freedom to be wrong:

'I've seen them get in an awful mess. They'll
persevere without asking for help.'

Not surprisingly, scripts in first year laboratories tend to
take more decisions upon themselves. Students often, like staff,
adopt the conventional wisdom that this is a necessary if un-
fortunate consequence of lack of maturity:

'In the first year... you were given a script and it said,
you are going to work out..., you will need to..., it
was routine.'

'It was necessary in the beginning, I think, but towards
the end we found we didn't need as much detail in the
script as we were given.'

'I don't see them getting away with it in the first year.
Then you really would be lost.'

GIVING DETAILED INSTRUCTIONS

Sometimes necessary practical instructions, such as how to
work a vacuum system, have the nature of giving necessary
background information (they do not pre-empt significant
decisions) while presenting all the difficulties inherent in giving
explicit practical directions in writing. So, for example, a
script can contain:

'Close valves 1 to 7. Open valves 2, 5, and 3 in that
order, and then close valve 2...'

continuing in the same vein for more than a page. Much the same can be true for instructions having a more physical content (those for circuits in electronics are an example). In such situations, it is not surprising to find students saying:

'If somebody says, do it this way, because maybe it's the easiest way - they know more than me, fair enough, but I'd like to be able to see why I'm doing things.'

If detailed instructions have to be given, a step by step listing seems likely to be clearer than continuous prose. Thus, for example:

Switch on the oscilloscope and signal generator.

Connect up the circuit shown.

Position the trace using the X SHIFT, Y SHIFT, and VERNIER controls to give a trace with a single leading edge of a pulse.

Measure the rise time of the leading edge.

They are, of course, better suited to situations where the student must carry out a fair number of steps correctly in order to arrive at a point which demands the taking of thought, than they are to ones where the taking of thought is of the essence from the start. Indeed, their main point is to avoid overloading the memory with what will be irrelevant detail as soon as it is past.

However, at somewhere near the opposite pole from those who wish to leave most decisions, in student's hands, are those who wish to focus his attention in very definite, pre-planned ways. In its strongest form, this view converts the laboratory script into a programmed teaching manual.

The point of the programming strategy is usually to take attention from practical decisions, so as to focus it on ideas. Thus its use is often where practical work has a mainly illustrative function. Correspondingly, the steps in the instructions are now questions to answer or consider, for example:

Describe the appearance of a differentiated perfect square wave.

How does your trace resemble a differentiated square wave? Can you account for any differences?

Suggest conditions involving R and C which will ensure that the output from the network is to a good approximation the derivative of the input.

This is, of course, spoon feeding. The question is when and where it is appropriate. Good programmed inst ructions are not easy to write, and it is all too easy for them to fail to convey the larger significance that lies behind the steady flow of tiny steps. At the same time, they can be and have been used profitably in laboratories, often mixed with other more open activities.

GIVING BACKGROUND INFORMATION

Sometimes the script writer seems to see it as his job to produce a self-contained package, especially where the chance that a student may do an experiment before being lectured to on the theory is high.

However, others (see above) give only book references, so sending students to the library in laboratory time. It would seem reasonable to shift from the first to the second policy as time goes by, even in the end deliberately giving less than comprehensive references:

> 'The references had nothing directly to do with the experiment. They were all about concentric coils, and it wasn't, so we had to make up the theory for our-selves. It wasn't too difficult.'

This would seem to be an encouragement not to treat theoretical results mindlessly.

While some of the theory given in scripts sets out the ideas clearly and simply, there does seem to be some tendency to make it unnecessarily difficult, perhaps because of an attack of what has been called 'rigour mortis', brought on by the thought of colleagues reading it critically. Thus, for the first year students for whom it was written, the following does seem to be in danger of overstepping the mark:

> 'Let ϕ be a scalar quantity representing a wave travelling with velocity c... All such waves, no matter what their physical nature may be, are described by a single equation:
> $$\nabla^2 \phi = \frac{1}{c^2} \frac{\partial^2 \phi}{\partial t^2}$$
>
> $\nabla^2 \phi$ (pronounced "delsquared phi") in Cartesian co-ordinates is:

$$\frac{\partial^2 \phi}{\partial x^2} + \frac{\partial^2 \phi}{\partial y^2} + \frac{\partial^2 \phi}{\partial z^2}$$

The symbol ∂ stands for partial differentiation...'

all this being for an experiment in which standing wave ratios were inspected.

SETTING STANDARDS

Scripts often attempt something by way of setting standards; of indicating how far work should be taken, what things should be taken into account, what sort of analysis is expected, and so on. They do it both covertly and overtly. Overtly, for example,

> 'Some analysis of errors both systematic (e.g. radiation loss, velocity variation) and random is expected.'

indicates requirement and level. But rather little of most laboratory scripts is normally given over to any detailed spelling out of what is expected, not least because it is much harder to spell out such things than to describe how to connect a circuit.

Very often the job is done by a hint:

> 'Care must be taken in choosing a temperature co-efficient appropriate to the temperature range covered.'

which, while indicating a specific point, is little more than a flag indicating that something must be done about it.

Here is one important area where script and demonstrator functions are intertwined. If standards are communicated, it is likely that most of the work is done by demonstrators (see 3.3) and when work is marked (see 3.5). However, more perhaps could be done, maybe by way of sample reports with critical comments.

STYLE AND FORM

The written script relieves the demonstrator of the need to talk to students about at least some matters, relieving him of more the more it spells out.

One problem is that some things may be better said face to face than written, either because they should vary from one student to another, or because they need to be communicated with tact and care. A laboratory based largely on writing must appear rather static and unresponsive, because writing is like that. On the other hand, written material can give a feeling of certainty and clarity. Judging the balance is an important part of deciding the tone of the laboratory.

Finally, at the end of the day, the way things are said matters, and can convey important messages. The two sentences,

> '... the final result is one linear relation between the bending moments and the curvature of the longitudinal section...'

> 'The particle under observation is being knocked about by the thermal motion of the surrounding molecules.'

convey different ideas of what ought to be conceded or expected by way of full scientific formality (it is hard to find examples more relaxed than the second). It may be worth remembering that it can be easier to convey a consistent tone in discussion with students than in writing, especially when the writing is done by many hands.

This is not to say, however, that scripts cannot convey useful messages. For example, where others might baldly begin 'It can be shown that...', the following sets the experiment elegantly within the civilised culture of practical work:

> 'G. F. C. Searle was a notable figure in the teaching of experimental physics, inspiring awe, not to say fear, in generations of students... the present experiment, one of his best, is well worth reviving.'

3.5   MARKING REPORTS

There can be two broad kinds of marking: 'marking off', when a check is made that one experiment is finished and another can be assigned; and giving a mark to a full-blown report, which is recorded and counts for credit. In some places the two are not distinguished, but in many only a fraction of experiments are fully written up, in several cases some time after the experiment is done so as to encourage good daily record-keeping habits.

Practice varies on almost every dimension. In some places only post-graduates, and in others only staff, mark work. In some it is done in a face to face interview, in others in private with a hand-in hand-back system of trays, shelves or pigeon-holes. The time spent in a marking interview varies from five or ten minutes to an hour.

Where a few full write-ups are required, the marking off interview often serves to give pointers to matters needing attention in a full report.

There are several functions which marking can serve:

a regular check on work, and encouragement to keep at it.

a formal acknowledgement of good work, or an indication of poor work.

transmission of ideas about what constitutes good experimentation.

indications of how work ought to be written up.

development of relationships between staff and students.

feedback to staff about students and experiments.

discussion of the interpretation and analysis of results.

more general discussion of ideas in the subject.

In addition, marking can be seen by students as needing to satisfy several criteria:

as indicating 'what they want'.

as being fair or not.

as teaching something useful. .

as giving support and encouragement.

Such an analysis is not difficult to write, but it lacks substance in the absence of any concrete picture of what can really go on in marking sessions. Next, therefore, we offer some stories taken from observation in laboratories.

One such interview seemed to us to have a number of admirable features. There is no cause to conceal its origin; more of its context can be found in the description of East Anglia in chapter 5, section 5.5.

The demonstrator was brisk and open, and the student confident. Thus, probing the theory written out in the report, the demonstrator openly gave the reason:

D   What I'm asking is if you copied this without understanding it.

The student had used a computer programme to do a least squares fit, and was asked what the errors on the parameters were, replying that the programme did not give them. He was told that there was one which did, and that he should locate it. The student had tried writing his own programme to attach integers correctly to measurements taken from fringes:

D   Where did that get you?
S   It didn't get me very far.
D   I can see that.
S   I said, perhaps I'm doing this wrong - so I made up another programme.
D   Have you done any physics in all this?
S   It's all physics!
D   What a difficult way of doing it! You've done a lot of hard work, and I'm still not sure you've got co-incidences.

Then they looked together at a long table of values:

D   You don't believe all those!
S   Up to about five I do.
D   Why five?
S   The sheet says you should get about five.
D   That's no reason! You're at it again. The computer gives all sorts of numbers, and you have to decide as a physicist which to believe. Don't let me see you do that again.

adding, in mitigation,

D   I've got to pick on something - with you there's not much to pick on.

At the end, as was his usual policy, the demonstrator asked the student what he thought the report was worth:

D   So what do you think? Is it as good as the others (you have done)?
S   It's all in here (reads from the report) 'The most interesting so far'.
D   How did its write-up compare?
S   I liked the first one, but I lost interest in susceptibility, but this was interesting.
D   There's justice done to this one - there's thinking about it, and a fairly advanced ability to put it down and analyse. What would you pick on as good?

S Using the computer, though I think I could have done it better.

D Do you think it's worthy of either of the previous two?

S I hope it's better than the last one, but whether it's better than the first one is up to you. If I say more you'll give me two out of ten.

D Yes, well there's lots of non standard stuff here - I'll give you $8\frac{1}{2}$ if I can come back and take the $\frac{1}{2}$ off if I ever find you not quoting errors again.

One important feature here is the clear and open acceptance on both sides of what is what. The work is both a piece of physics and a piece of work to be marked. Demonstrator and student shift easily from one point of view to the other (the student's efforts to write a programme, to discussing the mark). The demonstrator is clear about standards ('You have to decide ...') in a way which is definite but also puts the work in a wider perspective ('... as a physicist'). His judgement of quality is rather specific ('... there's thinking about it, and a fairly advanced ability to put it down and analyse.').

These things all connect with the way this member of staff sees his lab:

> 'If at the end they need me, I've failed. I say to them, It's your work, not mine.'

It is probably in the detail of such an interview that the student learns how the demonstrator thinks the report should be written up. In some places, general advice is given. In one, one demonstrator provides sample reports. Others give notes specifying what the report should contain. One gives a number of references, which include Trelease's How to Write Scientific and Technical Papers, F.H. Lucas' Style, and Fowler's Modern English Usage. The advice to write as if for a scientific paper sounds convincing to many staff, but does not always communicate enough to students. The student who said,

> 'It's got to be written out formally. You have to use the tenses - give the details: apparatus, procedure, results. There's got to be an abstract, like a scientific report.'

seems to have acquired an image in which work expected at school is muddled up with some new ideas. Another touched on an element of unrealism in the specification:

> 'They want you to say what problems you had, but in a paper you wouldn't say your mistakes.'

60

The member of staff described above was rather clear about what he wanted. As one might expect, post-graduate demonstrators are sometimes less clear. In a second year laboratory in which post-graduate demonstrators shared marking with staff (but did the bulk of it), one of them had several doubts:

> 'I find it hard to know what to look for.'

and pressed, said he was looking for,

> 'Anything, so long as it's methodical.'

He said that there was in the laboratory no discussion of marks, and that he did not know if marks were later adjusted to allow for variation (though in fact the laboratory handout said that they were). He did not think training was possible; that one had to pick up the idea of what to do from talking around on the job. He had worries about fairness, recalling a case where, having thought of the mark, he found from the record card that he had previously given that student a higher mark for what he recalled as worse work, and said,

> 'When you're a student, you think it should be fair, but when you start teaching you realise it's so arbitrary.'

He did, however, think of the value of a marking-off inter-view as pulling the work apart so that if a student came to do a full write-up he knew what to put in. After one such inter-view, he invited criticism from the two students interviewed. One revealed that not all post-graduate markers were alike:

> 'You really put us through it. But the last one; he didn't know what it was all about... you could baffle him - twist him round your little finger.'

In the interview, which concerned an experiment in which a coarse diffraction grating was ruled on a simple ruling engine of rather good kinematic design, and then used to measure and if possible resolve lines in the mercury spectrum, he covered in an hour all the main areas of concern, giving roughly equal times to each.

It is often said that post-graduate students are of most value when they have learned about the laboratory by passing through it as students. This was a rather extreme case, in that the demonstrator referred during the interview to his own three-year old laboratory notebook to see what should have been done in the experiment. In another place, the member of staff in charge explained that the current influx of post-graduate students from other universities was beginning to be a problem. Clearly, such a view can only be held where the laboratory has been stable for some years; it was noticeable that the places where there was some training of demonstrators were those where the laboratory had been recently changed, or where the university was itself fairly new.

Many post-graduate demonstrators, then, have to mark work with a highly implicit set of standards in mind. This is not to say that they do not have a clear idea of what should come out of experiments. In the case just mentioned (the grating experiment) the demonstrator was clear that kinematic design was such an issue, and brought that out clearly:

D   Did you read the references on design?
S 1 Yes.
D   Did you understand them?
S 1 Up to a certain point.
D   What didn't you understand?
S 1 About the bearings.
S 2 It seemed irrelevant - I didn't pursue it.
D   It isn't irrelevant. What this experiment is saying is that if you think carefully about the design of an experiment - and you'll have to for your project - you can do a lot better than you think, even with simple apparatus.

He made an attempt to teach a point about design, showing how hard it is to think of questions that bridge ordinary experience and formal knowledge:

D   (after a long pause) Why is a normal table with four legs unstable?
S   If the floor isn't flat.
D   (trying again) If you want a table that is perfectly stable, how many legs should it have?
S   Three, but it's not as stable really.

In yet another laboratory, post-graduates marked reports, but on a hand-in hand-back system. One expressed its limitations:

'Officially you're supposed to discuss the report with the students, but actually you put it back in the basket.'

He too felt uncertain about standards, had never been given any instructions about marking, and did not know what happened to marks or whether they were adjusted. He thought that he might be over-lenient, having never (this year) given the lowest possible mark, having given the next lowest only once, while having given the highest three or four times. Like others, he said,

> 'I feel sometimes I don't know what I'm looking for.'

and thought students had wrong ideas about it too:

> 'They feel that if it goes wrong they'll get a bad mark, but I like to see a bit of thought.'

Observation of interviews suggests the possible existence of a marking syndrome which might be called, 'What shall I pick on?' It has two aspects. First, the number of points picked on tends to be much the same for each student, with the number being made up by finer and finer points the better the work. Second, the report is treated as an occasion to find detailed things to criticise, rather than as an occasion for any general, especially favourable, comment.

For example, one member of staff in a first year laboratory was handing back a previously marked report. He said that the work was good, but not concise enough, and then that they would look at one or two small points. In fact, there were ten of them, ranging from criticism of the notation used, to challenging an improbable value of chi-squared given by a computer programme (which turned out to have an error in it). Others were criticism of the student's use of the concept of best fit, arguing that any line can fit any set of points, and that the need is to show that the data are consistent with the fitted line; and a criticism of the way the independence of one variable was claimed from theory, not from experiment.

Strongly represented amongst the points raised, here and in many marking interviews, is error analysis. This is very understandable, but may not solely derive from the importance of the topic. It could also be an unintended outcome of the laboratory situation. Where the script fixes the method, where the theory and the result are well known, the analysis of errors is almost the only area of choice left to the student, so that inspection of what he has done in that area is the only remaining way to test the quality of the work.

It is not necessary to argue that errors are unimportant, to maintain that other things are important too. Indeed, it was noticeable that in two second year laboratories where scripts left much more of the method and approach to be decided by the student, the focus in interviews was less on errors and more on choice of method and experimental tactics.

63

In conclusion, a number of general issues seem worth raising. One is the point, noted earlier, that there appears to be no belief that marking is a discussable activity, or that demonstrators could be taught to do it better. It is seen as coming out of the person's expertise as a scientist, and his implicit knowledge of the purposes of the laboratory. Accommodated in this view is the traditional independence of university teachers, so that individual variation is seen as both inevitable and proper. Students do not fail to take the last point:

'It all depends on who you get.'

Clearly, the marking interview is a teaching opportunity. Those who operate a hand-in hand-back system miss a chance, and it was noticeable that in one laboratory where reports were handled in this way but where staff said that they liked to discuss reports face to face as well, they also complained that students tended to take the reports off the shelves when they were not around, so as to avoid talking about them.

Those who take the chance may not always get from it all that they wish. One effect can be to dampen down interactions between staff and students in the laboratory - perhaps the greatest amount of talk and discussion we saw was in a laboratory where it was policy not to have any assessment for the first part of the first year. Some try to distinguish the marking and helping functions, perhaps by having staff mark and post-graduates just help with experiments. This can - though it need not - trivialise the post-graduate's job, restricting his interaction with students to minor ad hoc assistance.

The marking discussion is necessarily an ambiguous event. It cannot be pretended that it is purely a discussion of physics, but nor is it satisfying to treat as a purely pedagogic affair, itemising flaws in work and presentation. So a question like, 'What is the error on this value?' works at two levels: in one sense as a question one physicist might ask of another, and at another level as the question, 'Did you work it out as you are supposed to?'. A good, if bizarre, way to grasp what it may be like for the student is to recall an interview for a job. There, every question is, on the surface, an interested inquiry, but underneath one knows that it is intended to provoke a reply which will show the questioner the sort of stuff one is made of. The inner conflict in such situations is a real one.

3.6   LEARNING WHAT IS EXPECTED

Just what is expected of a student is often couched only in the
most general terms.   This is partly because it is felt to be
that kind of knowledge which is absorbed slowly and not the
kind which is told directly, partly because being precise about
it is difficult and may mislead, and partly because it is the
kind of common knowledge that to the experienced soon comes
to seem obvious and not in need of explanation.

It was noticeable that staff, asked about what their
laboratories were for, were hesitant and tended to be unsure.
Remarks were often prefaced by, 'I think...'.

Students pick up clues about what is expected in many ways.

They soon notice what different members of staff expect:

S 1   (speaking in quotes) 'Mr Y will help you with your
       errors'.
S 2   He's a one for errors.   Mind you, if you wanted to
       know anything about errors, you should go to him and
       ask.
S 1   You soon find out.   Just you... have your book marked
       once by him, and that's it!   You're there for three
       weeks trying to work them out, and still getting it
       wrong.

Another student illustrates the subtlety of their perceptions
of how messages are transmitted.   She had just completed her
first attempt at a least squares fit, and, asked why she had
done it, said that it looked better and was anyway more accurate
than drawing a line by eye.   Asked if anyone would grumble if
she did not she explained,

> 'We've just done it in statistics, so I reckon they're
> looking for us to use it.'

adding when asked what they might say if she did not use the
method,

> 'They'd ask you, "Have you perhaps considered another
> method?" '

Naturally, much of what students pick up is from the marking of reports, and they attach some value to marks as a
communication system:

> 'Mine came back with lots of red on it, but no mark.

You like to know how you have done.'

'Mine had just one comment - "Why have you left a blank page?" '

Another student had picked up a very clear idea of 'what they want'. Asked how he would select the experiment to be fully written up, he said,

'The one with the prettiest error analysis. That's what the marker seems to want.'

Marks obtained of course tell the student how well he is doing, but are at the same time open to doubt, about fairness and about what is looked for:

'The last two (reports) were marked savagely. The first was my fault, but maybe the second was prejudiced It's hard to know what to put in. I had a look at some other people's, and we are given a sheet, but it's not much help.

Included in this is an uncertainty about whether what is valued for marks is the report, or the work done, even when the importance of both is acknowledged:

'You are marked on the report, not on the experiment. You could do a lousy experiment and a good report. At the same time it is important - we have to learn how to write them.'

Many clues are picked up, it seems, from small points which can be generalised. Thus, one pair were asked how they had obtained the mass of a flywheel:

S Well, we found these numbers on it - it's quoted in pounds.
D No, that's cheating, I think.

Others are put by the demonstrator in a more general light:

S There's an error there which doesn't correspond to the rest.
D OK. When you get this sort of thing you should do that one again.
S Yes, but once you change the speed you can't get it back to the same again - this coarse control isn't sensitive enough.
D OK, you've got a problem there, it's true. But you

can't go ignoring - I mean, this is how discoveries in physics are often made.

One problem students have is deciding how far to go. Enthusiasm suggests pursuing a host of ideas; duty requires the pursuit of an unknown minimum number; self-interest indicates a compromise. Sometimes staff give specific advice:

D   It's up to you how far you want to go (with a computer simulation)

S   Well, I could bring (the simulated satellite) back to Earth.

D   Maybe your time would be better spent. There's no end to these things... you can go on for ever... you'd get more credit doing something else.

However, the overwhelming impression from talking to students and staff in laboratories was that students learn what is expected, not mainly from specific advice, but by picking up a host of individual clues about all manner of things.

Everything in the laboratory contributes: the nature of experiments; the kind of help that is given or withheld; what other students say; and perhaps most of all what is said when experiments are checked or marked after being done. From these messages, specific and general, clear or unclear, they make inferences. Some are detailed, clearly understood, and worth learning:

S 1   He was obviously looking for the errors.

I   Did you learn much from that discussion?

S 1   I think it taught us that we have to be more particular when we write up, and follow every oddity in the results...

S 2   And that when we use errors, we don't use them straight out of the book, but try to apply them to the situation of the experiment.

Other lessons are much more general, deriving from the whole structure of the laboratory:

S 1   ... we're free to take it out (of the apparatus) and throw it away and see whether it's better or not...

I   How do you know you're free?

S 2   We just assumed -

S 1   Nobody's going to stop you.

S 2   It's probably stated somewhere, but I wouldn't know where... They ask you why you did it when they're marking... As long as you can give a good explanation, they'll be quite happy.

These were the students quoted at the beginning of section 3.1 as distinguishing sharply between the freedom of the second year laboratory and the constraints of the first.   It is at least possible that the explanation staff and students offer, that first year students cannot be expected to show much initiative, is not the whole truth.   The system itself maybe as powerful a driving force as any supposed inherent qualities of different groups of students.

A contrasting discussion comes from a first year laboratory, in which several experiments were cast in a programmed form:

I   I was impressed by how much time people spent writing and reading.   Did you do that, would you say?

S 1   Yes...   I was trying to keep up, going along.

I   ... what sort of things are you doing?

S 1   Reading up the instructions about what we're supposed to be doing...   then you make the measurements...   then write up after.

S 2   We're supposed to keep a fairly good record...   but we haven't been given any firm rules on it - you must do this and not that - most of us seem to write up as much as we can...

I   ... do you take your books to be looked at by anybody?

S 1   They're taken in to be marked - I don't know when - I think it's at the end of this term.

Here a quite different equally self-consistent system generates different messages and different uncertainties.   The existence of detailed instructions seems to lead to work being seen as 'keeping up', and to it being seen as a laid down sequence. This may not be wholly intended, as the following exchange hints:

S   Do we have to...?

D   We would like you to...

Because instructions are programmed, staff have less need to look at notebooks, leading to an absence of messages from marking discussions, so that students may tend to play safe and, 'to write up as much as we can'.   This may or may not be the intention - here it was not.   Where work is marked, students speak of 'what they want' as at least guessable; here they mention the absence of firm rules.   The deferring of assessment can have a number of valuable consequences, but can, in the absence of much interaction between staff and students, have unintended side effects.

Overall, then, the picture is one of learning what is expected as one goes along, more from guessing the meaning of particular instances, than from explicit statements. Indeed, students are inclined to be sceptical of policy statements, and to decide how to interpret them in the light of the practicalities of the situation:

S 1 It's not catastrophic if you do six experiments instead of eight, I don't think.

S 2 But there's always a feeling, well I ought to do eight.

Even though they may sometimes be unintended or inconsistent, it is the messages delivered by the actual situation to which students pay most attention.

4. Parallels

4.1 TRENDS IN TEACHING LABORATORIES

This chapter discusses published accounts of teaching laboratories, and tries to fit them into a general framework within which they can be compared, and trends discerned.

In simple terms, several tendencies are easy to see. Much of the published work is concerned with moving from what is often called 'cookbook' experimenting, towards something which is usually called more 'open ended'. However, not a little takes what might be thought the opposite direction, turning experiments into what are hoped to be clearer and simpler illustrations of ideas or principles. That is, experiments become what are often called demonstrations.

Alongside these trends is an attempt to design what the student does in the laboratory more carefully, in accord with some analysis or other of what it is hoped he will learn. This takes two forms: designing work to teach particular techniques; and designing work to bring out rather exactly defined aspects of what some call 'scientific method' and others call 'the art of experimentation'.

Finally, there is a tendency to look again at the organisation of the laboratory, whether this takes the form of making it into a kind of library of permanently available experiments, or of breaking the traditional pattern of isolated work at individual stations, perhaps by working in groups or by using seminar methods.

All such changes tend to be seen against a familiar and well-understood pattern: the 'traditional teaching laboratory'. It is widely assumed to present problems. One is the scarcity of evidence that it achieves what it purports to achieve. Another is a feeling that neither students nor staff get much pleasure from it - 'routine' is a frequently used term. Taken together with the visible expense, in manpower and hardware, of the laboratory, the literature exhibits a considerable sense of depression with the traditional laboratory.

Published accounts of laboratory innovations are, however, almost certainly not a good guide to what exists, nor to the general pattern of change, for several reasons. Laboratory teaching is part of normal work, and many teachers would see no reason to publish accounts of what they have done in what they see as the normal course of duty, even if there existed recognised channels of publication. Further, that which is published is very likely to be that which can be described in clear, perhaps dramatic ways. Many changes, of course, are complex, having reasons as much to do with local arrangements as to do with any simple and clear-cut analysis; they are at once hard to describe and hardly seem worth the bother. They may not have any clearly rationalised basis, being largely intuitive and pragmatic; not that they are likely to be the worse for that.

In this chapter, then, published accounts are not looked at purely in their own terms. Rather, an attempt is made to find ways of drawing parallels between them, and between them and what experience suggests goes on in many others; parallels which may help clearer thinking about laboratory work in general.

4.2 PROBLEMS OF DESCRIBING LABORATORIES

At first sight it seems simple to describe a laboratory: one just says what the experiments are, and what the students have to do.

This will not do because in education, as in politics, words which seem merely to describe are in fact intended to have a persuasive power as well; to put what is described in a favourable or unfavourable light. Clearly 'cookbook' is a term of this kind, but so for that matter is, say, 'basic techniques'. For that reason, the chapter introduces several new descriptive terms so as to force reconsideration of the intention of some common existing ones.

It will also not do because in education, there is a sense in which one does not know what one is talking about, even when one thinks one does. In physics, events fall into place because there is an agreed theoretical framework. A capacitor is charged by being connected to a potential difference, for example. But in the early days of electricity, and analogously now in education, experimenters wrote of pieces of amber rubbed by fur acquiring the power to attract scraps of paper, and could not be sure what in the description was essential and what was not.

In describing teaching, the difficulty is compounded by the

fact that the teacher, unlike a piece of amber, has intentions in mind. So it is all to easy for him to suppose that what he deliberately did - perhaps giving students a programmed instruction manual - is the relevant factor which makes any important difference. Obviously though, it may not be; the real deciding factors may lie elsewhere.

The problem infects all subjects which lack a good body of theory. It is, for instance, clear to anyone who has answered a questionnaire that the natural assumption that each person has answered the same questions is false, even though all had the same questions on paper.

These difficulties are cause for caution, not for despair. What can be done is to try to describe laboratories in an interesting and illuminating way; that is, to bring out possible connections or ways of looking at things which strike one both as previously unseen and yet as in accord with common experience. Those who have worked in teaching laboratories know a lot about them, and a good way of looking at laboratories is one that makes them want to say, 'Yes of course, though I never thought of it like that before'.

With these cautions, and the further caution not to expect any quite remarkable or stunning insights, we proceed with a little botanising amongst the laboratory literature.

4.3   'OPENING UP' PRACTICAL WORK

Menzie has entertainingly traced the sources of one kind of dissatisfaction with laboratory work, its static content and deadening form (Menzie 1970). As to content, he shows how recognisable today are many of the list of experiments known as the 'Harvard forty' proposed in 1886 as a sound basis of experimental work required of students entering Harvard. The question of changing content will be taken up later. As to form, he suggests that what began as work done by students in small numbers under the direct tutelage of the best scientists, became with the enormous growth of higher education, something more like a standardised assembly line of experiences supervised by teaching assistants whose instructions did not extend to deciding what might be best for the individual student.

Menzie traces growing dissatisfaction with this dead science, a dissatisfaction which has helped the rapid growth of project work in final year laboratories, documented by Chambers (Chambers 1970) , and discussed in chapter 7. One of the contributors to the present book recently essayed a tongue-in-cheek characterisation of some laboratory work as a formal social ritual (Ogborn 1976).

'Opening up' has, however, been attempted in earlier years too. Shonle describes one such attempt in an introductory laboratory (Shonle 1970). To be compared with it, from several points of view, is a variant on project work for the senior year in an American State university, described by Graetzer (Graetzer 1972).

In Shonle's laboratory the work done with a particular set of apparatus is divided into two phases, which he calls 'exercise' and 'experiment'.

In the exercise phase, the student is given a definite simple thing to do or measure, for example the focal lengths of two lenses (positive and negative), the acceleration of a vehicle on a tilted air track, or the wavelengths of lines in the hydrogen spectrum. Hints are provided.

In the second phase, the student decides for himself what to do with the apparatus used in the first. Some might repeat the exercise with greater precision. More would choose some new aspect to look at: in the case of the air track these can include momentum conservation, energy conservation, free, driven, or coupled oscillations, motion observed in a moving frame, and so on. As Shonle emphasises, it is clearly important to have flexible, multi-purpose apparatus, of which the air track is again a good example.

Graetzer's laboratory is also organised around apparatus which can be used in many ways. The students begin a project with each group of apparatus, but instead of continuing, move after about every three weeks to another group of apparatus. There they find the record of what was done by those who had it before them (having left their own for others), and pursue the project in their own way for another three weeks, after which they again move on.

For example, one pair who first used nuclear coincidence equipment tried to detect electron-positron annihilation, but only got as far as calibrating the detectors. The second pair repeated that work with narrower ranges of energy, and got coincidences, studying also the effects of delay in one channel. The next pair changed the direction of the work, and tried to measure the speed of the gamma rays emitted, but could not resolve the small time difference involved. A fourth pair looked for coincidences between gamma rays of different energies from

Co-60, but it took work by a fifth pair to detect them. The sixth pair exploited the previous work to set up a two-detector cosmic ray telescope, and another pair tried with only partial success to set up a Compton scattering experiment.

Obviously the leaving behind of a record of work done plays an essential part in the organisation of this laboratory, in which work on a project is cumulative.

In comparing the two laboratories, we need a term for an identifiable chunk of work with a beginning and an end. We will use the term 'task'. Thus Shonle's students have tasks falling into two parts, while Graetzer's have tasks which involve taking over and contributing more towards an experiment.

Shonle's own term 'exercise' seems a useful one to describe work of the kind given in the first phases of his tasks. We will use it for work which practices some definite skill or involves some small, definite, short term goal not chosen by the student, which is given to the student not as an end in itself but as a means to some other end. Thus an exercise compares with piano practice or with making a single special joint between two pieces of wood, as opposed to performing a sonata or building a complete cabinet. In this sense Graetzer's students do no exercises.

It is harder to describe the second phase of Shonle's tasks. 'Experiment', or 'project' are too diffuse to serve. The work in Graetzer's laboratory has long term and shifting goals, not all of which can be foreseen at the start, and for such tasks the term 'project' seems appropriate. By comparison, Shonle's students ring changes on given apparatus, within a fairly tight framework. Perhaps the term 'improvisation' (with its suggestion of a solo musical performance within set harmonic limits) might serve.

A number of other descriptions of laboratories involve 'opening up' practical work to the extent of giving opportunities to students to make such improvisations. The idea of an 'instrumented laboratory' described by Soules and Bennett may belong to this family (Soules and Bennett 1968). Like Shonle they are concerned to provide flexible multi-purpose equipment, but they put the emphasis on instruments and transducers of various kinds. King has also urged a similar emphasis (King 1966).

Clearly 'opening up' practical work means many different things, a fact well brought out in a report of a conference on the 'divergent' laboratory in which deep differences of opinion tended to be masked under the same words (Ivany and Parlett

1968). Some writers, like Shonle and Graetzer, have a definite strategy worked out (Beun 1971, Fox 1967). Fox reports tasks very much like the improvisations of Shonle, while Beun takes the perspective of the whole undergraduate course, with the gradual development of project-style work within it. Others work more on a principle of variety (Finegold and Hartley 1972, Finegold 1972, Aldridge and Feldker 1969). These writers have introduced into their laboratories alternative activities such as paper planning exercises, computational work, the development of a new laboratory experiment, and so on.

Not surprisingly, most authors strongly emphasise the time needed by students when work is made less routine. Some, in the American context, have found it possible and have felt it proper to open the laboratory throughout the working day, so as to remove the constraint of fixed hours.

Also, most authors have as a main aim encouraging students to exercise a wider range of critical skills, including thought, planning, and inventiveness, than they see as called forth by traditional set experiments. There is, however, some tension between this and the illustrative function of practical work. Notably, those like Shonle with whom students choose for themselves improvisations which could have been (but were not) set as experiments to illustrate a principle, stress the illustrative aspect more than those like Graetzer, for whom the processes of scientific inquiry are paramount. Some of course fall in between.

Finally, the choice does not simply lie between projects and set experiments, with projects perhaps reserved for the final year. Work in which students make some, but not all, of the decisions they need to make in a project where they start with nothing more than a long term, perhaps ill-defined, goal, can be devised. Such work can occupy times more like three weeks than three months. Improvisations on a theme with set apparatus are one possibility; small projects, design exercises, and cumulative experiments are others.

4.4   EXPERIMENT AS DEMONSTRATION

Most practical work has in it some element of witnessing nature working as theory says she does. One trend in the literature is towards stripping the conventional experiment of all aspects except this one; towards converting experiments into demonstrations the student performs for himself.

One reason given for such a shift, in introductory courses, is that other teaching cannot, mainly for lack of time but also for lack of chances to provide relevant experience, get students to think long and hard enough about concepts to understand them well (Goldberg 1973, Price and Brandt 1974, Long 1975).

A related aspect, mentioned by some writers, is the conversion of experiments into more or less self-service, self-instructional packages. The traditional experiment, with its script, its apparatus set up in a special place, and its tendency to be self-contained, has always had about it some trace of this flavour, but here we refer to a much more radical redesign, even to the point where the apparatus works in push-button mode (White et al 1966).

Belonging here, but also elsewhere, is the idea of the laboratory as library (Oppenheimer and Correll 1963, McCord 1968). That is, experiments are available off the shelf, sometimes at any hour as well, rather than being permanently set out. The concentrated nature of demonstrations, and the likely need for many more of them than of longer experiments, means that they naturally lend themselves to such a pattern.

Just what kind of task makes a suitable demonstration is not seen in the same way by all. Goldberg describes what amount to step by step teaching programmes, specifically designed to teach concepts. So, for example, a student is set to record the path and velocity of a frictionless puck as it is scattered by interaction with a fixed one, to compute angular momenta for it and test for conservation, and then to do the same for its kinetic energy, testing whether the force field is conservative. All this might describe a conventional experiment, but Goldberg's have been shorn of technical difficulties, of decisions to be made by the student, and of error analysis.

Price and Brandt, in common with Goldberg, see complex apparatus or practical difficulties which others might regard as interesting, even as the essence of the matter, as distractions from the job of presenting an idea in action as clearly as possible. Their laboratory is adapted from one described by King as a 'corridor laboratory'; that is, a free access exhibition (King 1968).It is, however, not an exhibition; rather, it contains ready set up apparatus with which the student can get results in about a quarter of an hour, together with photographs, graphs, books, and instructions, all designed to make the experiment largely self-instructional and self-contained. Some of the instructions are tape recorded, and the recordings also include teaching about the theory and its relation to the experiment.

By contrast, others have developed the exhibition aspect (White et al 1966). Their experiments are designed to be as much like science museum exhibits as possible, though always with a quantitative side. The apparatus is contrived to go through a pre-determined routine, usually operated by push-buttons or the like, and to display the relevant physical behaviour clearly and attractively in a qualitative way, besides yielding data.

Perhaps the essential feature of all such work is that each experimental task is 'packaged'. That is, it comes complete with everything that is needed, and is limited to a fairly short time and to fairly fixed purposes.

Several writers point out that, just as a package of soap is the size it needs to be to hold a bar or two of soap, so it is natural for packaged experiments to occupy the time they seem to need, as opposed to occupying a period defined by a time-table slot. Thus times vary widely, and in particular can be much shorter (perhaps half an hour) than any time which would usually be thought of as needed for an experiment.

It is, of course, not necessary to imagine the whole laboratory filled with such packages. Where they are used exclusively, it is often because very large numbers of students have to be handled. However, such packages can find a place in laboratories which contain tasks of many other kinds: one concerning the use of the oscilloscope being an obvious example. It is notable that it is in areas like electronics and computer simulation that packages have seemed to many teachers to be natural and acceptable. People to whom a demonstration task on, say, the viscosity of air at various pressures would be repugnant, have felt little difficulty in presenting, say, combinations of logic devices in this kind of form.

4.5 SINGLE-MINDED LABORATORIES

Laboratories devoted to a particular branch of physics are common enough. An innovation which makes a few appearances is to focus the laboratory, not on a topic, but on a particular kind of activity learned through several experiments. Thus, for example Goldberg (see above) is concerned just to get students thinking about ideas.

Richards offers a more radical variation (Richards 1974). He has a whole laboratory course devoted to developing what he calls 'experimentation', by which he means the identification, isolation, and control of variables so as systematically to discover functional relations between them.

An example may make the single-mindedness more apparent. A typical task gives the student a set of accurately machined nesting metal cylinders which are to be rolled down a very hard straight smooth slope. Means of timing the cylinders rolling down the slope easily, accurately, and repeatably, with minimum difficulty, are provided. The problem - the effect of mass and geometry (length, inner radius, outer radius) on rolling time - is well-defined. What the student has to do is to decide on purposeful, systematic ways to collect data in the light of some hypothesis, often obtained from dimensional analysis, about the underlying relationships.

Richards bases his laboratory on a definite, if perhaps restricted view of what constitutes experimentation. Nedelsky arrives at something not altogether dissimilar from an analysis of the educational problems of traditional practical work and of project work (Nedelsky 1958, Nedelsky 1965). He argues that the art of experimenting is hard to learn in conventional set experiments, because few if any decisions are left to the student, and because, as a beginner, the student cannot yet see his data as much more than sets of numbers without much meaning. Equally, he argues that loosely structured project work is difficult to supervise well, is expensive in time, and pays a cost in terms of limited sampling of content.

Nedelsky's solution is a format for tasks which clearly brings out the particular aspects of what is involved in experimenting that he has identified as his immediate goals. The task has an experiment-then-predict form: first the student investigates some phenomenon, with a view to becoming able to predict numerically some as yet unspecified piece of its behaviour, while knowing what kind of behaviour he will have to predict. For example, the student may be given some circuit components, and meters, and be told that later he will have to predict the reading of a meter in a circuit chosen by the teacher, together with limits on the precision of the prediction. In the investigating phase, the student can design what experiments he thinks fit, and is expected to collect data to be used later for prediction. (See 9.2.2)

The point of this format is to focus attention as sharply as possible on what is needed in designing experiments so as to be able to say what will happen in a given situation.

The above two examples have concerned 'experimentation'. Single-mindedness, with its possible values and limitations, need not of course be confined to that (see chapter 6).

The examples do suggest that it is worth distinguishing two kinds of 'exercise': an exercise in the sense of section 4.3 in which the main focus is on the physics involved, and 'methodological exercises' in the sense of the present section, in which the focus is on the process, and in which the physics content is seen as arbitrary. Of course, much practical work is intended to do both at once. It may be worth stressing the element of artifice often involved in methodological exercises; that is, that they often have about them a contrived air, because they are designed to bring out issues not related to the content, as clearly as possible.

4.6   INTEGRATION AND WORKING IN GROUPS

One of the commonly noted difficulties with practical work is its lack of integration with theory, when experiments are provided singly and are done in rotation. Those who try for greater integration are often led to working in groups, where apparatus cannot readily be replicated.

Others start from the positive advantages of group work and discussion methods, and again are led, sometimes but not always, to devise work which is more integrated. The two are then   related, but not in any necessary way.

It is notable that practical courses in electronics are often integrated (Babcock and Vignos 1973; see also chapters 5 and 6). The subject is seen as one in which theory and experiments can usefully and easily develop together; a point similar to that made in section 4.4, where it was noted that such subjects are often provided with theory and experimental work in a 'package'.

In other cases, greater integration is sought by some reorganisation of laboratory work. At its simplest, some have groups of students, briefed by or working with a demonstrator, working together on an experiment (Oppenheimer and Correll 1963, Barrett and Blake 1974, Fentem 1974, see also chapter 5). The experiments worked on in this way may be looked after by one demonstrator, so grouping experiments together (Duffin 1971, see also chapter   6).

The reorganisation may be more radical. King, for example, teaches a whole course in one concentrated period of a few weeks, working full time (King and Parlett 1971). In this pattern integration is natural, and in addition the group of students can not be very large, so some group work can be involved. When the practical work for a large group is divided into short units

79

each looked after by one teacher, it is possible for a unit to develop into a more integrated sequence of work, and for group discussion to be used.

Starting now from the other end, the use of seminars or discussions is reported by some. Tubbs describes laboratory seminars with small groups, concerning the best way to make a measurement (Tubbs 1968), while Mendoza has used large scale seminars with closed-circuit television to discuss the design of experiments and the interpretation of data (Mendoza 1972). Indeed, laboratory work, including projects, may be just one part of the work done in a group, which is also following a programme involving theory, examples classes, presenting papers, and so on (Black, Dyson, and O'Connor 1968).

Group discussion is yet another part of the attempt to loosen or 'open up' laboratories, as discussed before. It is a response to the criticism that while the laboratory ought to be a good place for teachers and students to talk and share ideas, it rather rarely is that in practice, except in a limited sense. The criticism may be taken to the point of seeing the laboratory as the place to develop skill and confidence in communication (as it usually is for just written communication)(See chapter 9). Group work can also be seen as a way to get more of what laboratories are often supposed to be about: learning by example from someone who knows what he is doing.

4.7   SUMMARY

From the literature, one gains an impression of changes in the direction of making experiments less routine and more investigatory; of attempting to clarify the main purpose of particular parts of the work and to design activities which more directly tackle that purpose; and of more flexible forms of organisation, whether of experiments being used in more varied ways, or of students working in more varied ways.

Observation and information supports the generality of the first trend, at least to the extent that very many physics (and other) departments use project work in the final year, and some in previous years. This, and the closer linking of laboratory teaching to research, is discussed more fully in chapter 7.

It is less clear how far any other trend can be detected in daily practice. Laboratory teaching is perhaps too flexible, individual, and adaptable for it to be likely that any one or two clear patterns would emerge. Change, however, there certainly is.

REFERENCES

(Fuller references for project work are given at the end of chapter 7.)

Aldridge W A, Feldker P (1969) 'The open physics laboratory' Physics Teacher, vol 7, pages 321-326.

Babcock L E, Vignos J H (1973) 'Op-amps as building bricks in an undergraduate project-type lab', American Journal of Physics, vol 41, pages 89-98.

Barrett J, Blake P (1974) 'The teaching of reaction kinetics to first year undergraduates', in Studies in laboratory innovation, Group for Research and Innovation in Higher Education, The Nuffield Foundation.

Beun J A (1971) 'Experiences with a free undergraduate laboratory', American Journal of Physics, vol 39, pages 1353-1356.

Black P J, Dyson N A, O'Connor D A (1968) 'Group studies', Physics Education, vol 3, no 6, page 289.

Chambers R G (1963) 'What use are practical classes?' Bulletin of the Institute of Physics, vol 14, pages 181-183.

Chambers R G (1964) 'A survey of laboratory teaching' Bulletin of the Institute of Physics, vol 15, pages 77-84.

Chambers R G (1970) 'Laboratory teaching in the United Kingdom', in New Trends in Physics Teaching vol II, UNESCO.

Duffin W J (1971) 'The undergraduate physics laboratory, Part II The grouped laboratory' Physics Education, vol 6, pages 144-148.

Fentem P H (1974) 'The integrated course in physiology and pharmacology in the first and second years of the medical curriculum at the University of Nottingham', in Studies in laboratory innovation, Group for Research and Innovation in Higher Education, The Nuffield Foundation.

Finegold L (1972) 'Open-ending a senior modern physics laboratory', American Journal of Physics, vol 40, pages 1383-1388.

Finegold L, Hartley C L (1972) 'An experiment on experiments in a senior laboratory', American Journal of Physics, vol 40, pages 28-32.

Fox J N (1967) 'Laboratory built on air', American Journal of Physics, vol 35, pages 789-791.

Goldberg H S (1973) 'An introductory mechanics laboratory at UICC', American Journal of Physics, vol 41, pages 1319-1327.

Graetzer H G (1972) 'Cumulative experiments in the advanced laboratory', American Journal of Physics, vol 40, pages 270-276.

Ivany J W G, Parlett M R (1968) 'The divergent laboratory', American Journal of Physics, vol 36, pages 1072-1080.

King J G (1966) 'On physics project laboratories', American Journal of Physics, vol 34, pages 1058-1062; also in New Trends in Physics Teaching vol II, UNESCO.

King J G (1968) 'Corridor-demonstration laboratory', Commission on College Physics Newsletter no 16, May 1968.

King J G, Parlett M R (1971) Concentrated study, Society for Research in Higher Education.

Long R (1975) 'Laboratory learning modules', American Journal of Physics, vol 43, page 340.

McCord W M (1968) 'Laboratory arrangement in a new perspective', American Journal of Physics, vol 36, pages 874-878.

Menzie J C (1970) 'The lost arts of experimental investigation', American Journal of Physics, vol 38, pages 1121-1127.

Nedelsky L (1958) 'Introductory physics laboratory', American Journal of Physics, vol 26, pages 51-59.

Nedelsky L (1965) Science teaching and testing, Harcourt Brace.

Ogborn J (1976) 'Etiquette in the laboratory', American Journal of Physics, vol 44, pages 625-627.

Oppenheimer F, Correll M (1964) 'A library of experiments', American Journal of Physics, vol 32, pages 220-225.

Price R M, Brandt D (1974) 'The walk-in laboratory', American Journal of Physics, vol 42, pages 126-130.

Richards M J (1974) 'A first year course in experimentation for mechanical engineers at Brunel University', in Studies in laboratory innovation, Group for Research and Innovation in Higher Education, The Nuffield Foundation.

Soules J A, Bennett R B (1968) 'The instrumented lab', American Journal of Physics, vol 36, pages 1068-1072.

Shonle J I (1970) 'A progress report on open-end laboratories', American Journal of Physics, vol 38, pages 450-456.

Tubbs M R (1968) 'Seminars in experimental physics', Physics Education, vol 3, page 189.

White H E, Weltin H, Gould M, Rice R A (1966) 'Quantitative demonstration exhibits and a new low cost physics laboratory', American Journal of Physics, vol 34, pages 660-664.

5. Places

5.1 INTRODUCTION

This chapter is written by people outside the institutions it is about. It consists of descriptions of the main activity in five university teaching laboratories - two first year and three second year - where the main activity is in the form of 'experiments'. The descriptions are not intended to be complete, but merely to convey a picture of the most interesting or character-istic aspects of each laboratory. They sometimes include an indication of changes by which the laboratory's present state was achieved, because such an indication can sometimes make the present state more intelligible, and because the changes can sometimes be as important as the present state.

The picture of each laboratory is selective. It is mixed with an element of comment, and factual information is often restricted to what is relevant to the comment. Important secondary activities, such as a short project or workshop course, are not necessarily mentioned. Nor are such important organisational matters as the numbers of people involved or the numbers of days of the week on which separate divisions of a year group come to the laboratory, unless they are of special interest. Statements like 'students work in pairs' are usually made with-out regard for the exceptions which, for various reasons, occur in practice.

The information on which the descriptions are based was gathered by groups of observers, each group consisting of three or four members most of whom were university physics teachers. Two or more sessions of each laboratory were visited by such a group, whose members observed and listened to the impressions of people working there. The visits were made in the academic year 1975-76.

5.1.1 DIFFERENCES BETWEEN LABORATORIES

There are necessarily important variations in the way the five teaching laboratories run, for two reasons: the difference in the experience of the students between their arrival at the university and the end of their second year, and the difference in the numbers of students in the various departments. Consequently there are quite wide variations in the meanings of words and phrases in the different parts of the chapter. In particular 'experiment' and 'script' must be remembered to be merely generic terms. If a pair of Sussex first year students suddenly changed experiments with a pair of second year Bristol students, all four of them would be horrified. The word 'demonstrator' has been used, not in its local senses, but to describe any graduate (other than a technician) employed to teach in the laboratory, in order to avoid explaining the word's use at each place. But the job of being a staff demonstrator or a post-graduate demonstrator, differs widely from place to place.

While this chapter is an attempt to display such qualitative differences, together with why they exist, it avoids quantitative comparisons from which the reader might be tempted to make his own deductions, because of the large additional body of information which would be needed to make such comparisons valid. To take an example, the proportion of the physics course's credit which arises from assessment of laboratory work could be given in the five cases, and would be different. But making those proportions comparable would require details of how much of a student's commitment is to physics, and then of how much of what might be termed maths, electronics, crystallography, and so on, is included in physics at each place, before even starting on considerations of course units and fractions of the academic year. But it is possible to state, upon this point, and the fact may be worth noting, that few students to whom we listened at any university could say what happens to any marks awarded for practical work, apart from their being written down somewhere.

The first laboratory described, that of the first year at University College, London, exhibits many features to be found in other laboratories. The description is therefore in some respects more factual and detailed, but less the subject of comment.

5.2   FIRST YEAR AT UNIVERSITY COLLEGE, LONDON

For most of their first year, the students at University College work on 'experiments'; the word is used in the sense of a specified task with an outcome forseeable at least in outline by good students. The experiments require equipment which is often purpose-built, bulky or expensive, and all students do not perform them in the same order but instead take turns at them; a common form of organisation which, in this book is given the name 'circus'. The main disadvantage of a circus is that each experiment is done by some students too early in their development, and by others too late; a disadvantage most acutely felt in the first year when the relative growth of students' capability is greatest. For this reason, University College uses two small circuses, one of simpler experiments in the first term and another of more complex experiments in the second. None of the laboratory work is related, except in the widest sense, to the lecture courses which students also attend.

In addition to the experiments, which are the main subject of this account, there are other activities. Outside laboratory time (three and a half hours on each of two afternoons a week) at the beginning of the first term there are about ten lectures on experimental errors and their statistics. Later that term two weeks of laboratory time are given to a course in Fortran programming. In the second term, a few at a time, students give two weeks of laboratory time to a workshop course. Two weeks near Easter are given to a 'mini project'. Three weeks in the summer are given to engineering design. Activities such as these, related to and supplementing the main laboratory work, are common in university laboratories.

Our visits to University College were made during the second term. About thirty students were doing experiments in a square room of about 250 m² and a dark room opening off it. A small library and technicians' rooms also open off the main room. The circus consists of about fifteen experiments of which each student would do altogether about six. Each experiment has a time, usually two afternoons, which a student is expected to devote to it. For some experiments there is only one set of apparatus but for others there are more, raising the chance of each student doing them. For at least one experiment, which all students are expected to do, there are four nearly identical sets. Students work singly, except at one or two experiments where a helper is needed to manage the apparatus.

Three demonstrators, a staff member and two post-graduates, were present all the afternoon, but these three people demonstrate only one afternoon each week, and another three demonstrate to the same students the other afternoon each week. The students are divided into three groups and each group has its own demonstrator each day, who assigns experiments, comments on results, deals with immediate problems, and knows and is known by his group of students. Immediate problems are also dealt with by any demonstrator available and by the technician.

5.2.1 APPARATUS AND EXPERIMENTS

The discrimination with which the apparatus has been assembled and the care with which it is maintained are notable. Much of it has been made specially for the laboratory, and such items have been thoroughly designed and generously made and finished. One student who, having been a lab steward at school, had become interested in equipment, said that of several universities he had visited, all with good equipment, University College was the best. A student coming from school would be expected to be impressed by the precision with which measurements might be made, and could be expected to infer from it an obligation to achieve accuracy in his results even if no such obligation had been stated. And appreciation of factors governing accuracy, together with its more certain achievement, are main aims of the laboratory.

For each experiment there is a 'script' which students read. The script tells students what to do, gives a background of theory, gives references to books, and gives occasional specific warnings and pieces of advice and information, in roughly this order of importance. Scripts for first term experiments are shorter like the experiments, and more instructional. Scripts for the second term are between two and seven pages long, and although in some cases the instructions about using the apparatus are quite specific, there are more in which they are given in general terms. The second term scripts often have several pages devoted to mathematical theory bearing on the experiments.

Some of the first term experiments are simply collections of exercises, the exercises relating to one another only in needing the same apparatus, usually common and important items like oscilloscopes and digital timers, in the use of which students need practice. These collections of exercises fit into the organisation as 'experiments', but 'exercises in techniques' describes them better. Some first-term experiments and most second-term experiments have simply definable end points - 'the electric field distribution ... is determined', 'the velocity of sound is measured', 'the refractive index of air is determined as a function of pressure', 'the variation of time period of a simple pendulum with the amplitude of the swing is investigated using a high precision timing technique' - these are taken from the short descriptions in the three-page list of second term experiments.

The scripts, apparatus, and verbal advice given by demonstrators and technician, imply that the resources of the
laboratory are directed towards enabling and helping students to
make their results accurate. For example, the 'velocity of
sound' script states that a precision of 0.1% can be approached,
and warns that the oscilloscope may impair this precision by
heating the air. The script does not, by reticence as to the
oscilloscope's heating effect, involve students either in finding
the cause of unexpected imprecision or in a larger spread of
readings which could lead to a discussion of errors in general.
Consideration of errors is, however, an important part of all
experiments, and the experiment which is most particularly
devoted to the treatment of errors is one which all students are
expected to do. Its ostensible purpose is determining the
viscosity of oil by dropping different sized steel balls through
it, and the measurements are all simple. But the estimation
of the spread of results would require too much calculation if
a computer program were not provided. The experiment's real
purpose, 'to illustrate the use of statistical techniques in data
analysis', is made quite clear.

5.2.2   DEMONSTRATORS

Students keep note books in which they record readings and
calculations. Towards the end of each experiment a student is
assigned to his next by his demonstrator on the basis of what
is free, what is suitably easy or difficult (for attempts are made
to identify and cater for weaker students), and the student's
preference. After he has started his next experiment the
demonstrator looks at and comments briefly and informally on
his result for the earlier experiment. Twice in the term he
writes up an experiment in detail, and this work is looked at
by a staff demonstrator and discussed at length when it is
returned.

The director of the laboratory had been in charge for several
years, and during this period he had made many adjustments to
its running. The laboratory had not undergone any single major
change but most experiments had been replaced at one time or
another, and there had been unobtrusive changes to such things
as the number of demonstrators, their role in assessment,
students' choice of experiments, students' recording of their
work, and so on. The changes exploited new possibilities and
overcame new problems as they arose.

The two post-graduate demonstrators whom we saw had themselves been students in the laboratory and had demonstrated in

it in previous years. Their familiarity with the physics of the
laboratory was a factor enabling them to look after all the
experiments for a limited group of students, an arrangement
which enables students' individual capabilities to be more quickly
known than one in which each demonstrator helps any student at
a few selected experiments. One demonstrator described his
job as being chiefly to be available to be asked questions arising
from the inexpertness of students and the failure of apparatus,
and neither demonstrator spoke much to students without first
being spoken to. Not being concerned with assessment of
students' work probably made it easier for the post-graduate
demonstrators to become aware of students' reactions, as, for
example, in the remark of one post-graduate, 'If asked to see'
(that is, 'if students are referred to',) 'a normal book, they
get frightened.' One post-graduate demonstrator, who took
students to see his own research, decided with care the time
in the year when it would be most profitable to do so.

5.2.3   THE STUDENTS' VIEW

The laboratory at University College does not involve students,
at the beginning of their university careers, in abrupt reappraisal
of the subject they have chosen to study. To them, physics at
the university does not seem to differ from what they would have
expected, in the qualities which distinguish it from other subjects,
although it seems more demanding and more rewarding than
before. A good student, asked about the difference between first
and second terms, said to us, 'The gentle start is a good idea.
I felt at first it was too like A levels, with the short one-after-
noon practicals - I was not sure it was good then - thought it
too easy. Now I see that if I had had to do the second term
experiments straight off at first, I would have been demoralised.'
This change of attitude is to do with growing familiarity and
confidence - earlier he had said, 'You hurry first time, but now
you keep your cool, then there's time to sort out most things.'

The comparative sophistication of the apparatus is appreciated
by students, who are enabled by it to get more accurate results.
When one told us, of one experiment, '... but once it's done,
the sheer elation to get 9.81 - that was good for days and days',
his neighbour countered that 'If the apparatus was good, you
would get good results'. The occasional failures which go with
sophistication are accepted as unavoidable, and the expertise
which surmounts them is respected. We saw a post-graduate
and a staff demonstrator struggle in vain with one experiment,
and heard the staff demonstrator greet the technician as he
arrived with, 'Do you have any magic method for putting this

right?' The technician soon won. To us he ascribed the frequency of his being asked for help partly to his being always
present, and becoming more quickly known to students at the
start of each year, whereas demonstrators are only in the
laboratory on alternate afternoons.

A quotation from the student who had been elated 'to get
9.81' summarises by implication much in this laboratory:

> 'The staff are very good. If you can't see where a
> wire goes, you feel an awful fool, but they never say,
> "You idiot". The technician is helpful too. For
> example I had a switch on the bench I didn't know what
> to do with. He puts you right. I would have spent
> half the afternoon looking for how to use it.'

He notes the patience. He shows that the difficulty of the
tasks has been set to be within reasonable ambition but beyond
easy accomplishment. He concedes, perhaps without realising
that there is anything to concede, that the planning of his work
is still being done for him.

5.3   FIRST YEAR AT SUSSEX

Sussex University is about twelve years old. Originally the
policy of not having students specialise too early resulted in
nearly four hundred students reading some physics (the preliminary course in the Structure and Properties of Matter) for the
first two terms of their first year. Of these four hundred,
most gave up physics after two terms in order to concentrate
on other subjects, and about one fifth became physics specialists.
A first year physics laboratory was therefore set up which was
in nearly constant operation from Monday morning to Friday
afternoon, allowing each student about three hours to do one
experiment every two weeks. Development of other departments
has now roughly halved the total number of students who come
to the laboratory, which is therefore now less constrained, and
can allow more flexibility in its approach to different sorts of
students, but the framework is unchanged.

It is called First Year Atomic Physics Laboratory and the
introductory notes for students state: 'The main purpose of this
laboratory is to provide a context for the discussion of some of
the physical concepts in the Structure and Properties of Matter
preliminary course'. Two communicating rooms about 10m by
15m are used. At any one time a room might contain four or
five similar sets of apparatus for each of four or five different

experiments. For each session the students are directed, by means of a plan put up at the beginning of the term, to an experiment, or to a group of experiments from which they choose one. The individual experiments of such a group are often purposely made to differ in the degree of difficulty. Although students are usually allowed their own choice of experiment within the group of experiments to which they have been assigned their preference is sometimes over-ridden for educational or logistic reasons at the start of the session concerned. The students work in pairs, and the plan is made so that students of particular subjects and in particular tutorial groups remain together.

The experiments are usually pieces of established physics, often with equipment designed for a single specialised use. For example, at the time of our visits one room contained 'Millikan' experiments, 'Franck-Hertz' experiments, measurements of the photo-electric stopping voltage at different frequencies, and spectrometer experiments. The method of each is prescribed by the staff who have arranged that all necessary apparatus will be on the bench. Students find out what to do both by reading a script of five to ten pages which can be taken away at any time, and by being told by a demonstrator at the beginning of the session.

The staff do not demonstrate, except that one member is present at each session to give any necessary overall direction, and he is often free to help students. The demonstrators are research students. One demonstrator is assigned to each experiment for the term, and he is given a training session and a folder of extra information on it beforehand. The student to demonstrator ratio is smaller than in most laboratories with the intention that the demonstrators should be able to give more positive instruction. The demonstrators assess students' work by looking at their note books which are handed in a week after each experiment is done.

The nominal time each student spends in this laboratory is only twenty-six hours; less than two thirds of the time in one five week unit laboratory at Birmingham (6.4, 8.4). The aims are therefore modest and the methods simple. Useful pieces of physics and ways of measuring are shown, so that students may become acquainted with them, using the apparatus and giving as much thought as they like to it. But the student encounters hardly any uncertainty as to what he is to do. For this uncertainty, however valuable the opportunities and obligations it gives to initiate and plan, would use a lot of time, and furthermore it would be hard to provide uncertainties at the proper levels for all of a very heterogeneous class of students.

5.3.1   THE START OF A SESSION

Our visits were made on days when most students were intending
physics specialists, and were towards the end of the second term.
Students arrived on or before time, and during the first half
hour of the session they seemed to us to do very little without
discussion with demonstrators.   A student, asked what demonstrators did at the beginnings of sessions, said: 'He might show
a film relevant to practical, as in the case of the probability
one (on errors, the only experiment unrelated to the lecture
course), or he might just have a little talk about it first and
write a few diagrams on the board, or he might just say 'Get
on with it' and hardly say anything, or they might not even see
you to start off with.   They might come round once you've
already been looking at the apparatus to see if you've already
started yourself'.   Another student, asked about the start of a
particular experiment, said: 'The demonstrator came round to
explain the apparatus as it was particularly difficult and there
is a lot of it, speak about various novelties in it ... pointed
out what all the galvanometers are for and told us a bit of the
theory, and let us get on with it.'

For the next hour the students assiduously used their
apparatus, so that there was more than an even chance of a
student's having his hands upon it if he were looked at.   Hardly
any time was spent wondering what to do or reading scripts.
Students, who were asked how they found out what to do, said
that the scripts were the most important source and that the
demonstrators were useful when the scripts were not clear.
The scripts are quite specific about the use to be made of
apparatus, as well as giving reference to text books and giving
some explanations of theory.

More than an hour from the end, students started to pack
up and go, intending to write up their note books at home.   The
policy for these written reports, well known both by demonstrators and students, is that they are to be written as if to
the demonstrator and need contain nothing that the demonstrator
would know already.   Most scripts list what the reports of their
experiments should contain.   Such a list specifies graphs to be
plotted and quantities to be calculated and also usually asks for
comments on some of these and for answers to questions which
are in the script.

5.3.2   STUDENTS' REACTIONS

The last student to leave, on one of our two visits, had done his experiment (on the stopping voltage of a vacuum photocell for light of various frequencies) on a previous day but had become interested in its deviation from text book simplicity and had come in for an extra session to try, by altering the routine, to understand it better. In addition to time with the apparatus he spent at least an hour discussing the complications with two demonstrators, the staff member present, and one of HELP's observers. One of the two demonstrators spent so long himself investigating these complications at a spare set of apparatus that we mistook him for a student. This work gave a strong impression of experiments well chosen for the interest they would excite and the absence of a convention, sometimes strongly felt elsewhere, that the laboratory is an experience to be survived without involvement.

Other students, asked in effect if they yearned for freedom to display initiative and felt cramped by the closeness of the direction, denied both and seemed to think it quite natural that they should be directed to the extent they had been. Two demonstrators spoke of the main anxiety apparently felt by students as being to finish and go. One said: 'It's difficult to get students to enjoy what they do', and another about one experiment: 'They aren't interested in the experiment'.

A lot of care has been given to training the demonstrators for this course, because of the large number of demonstrators needed and the lack, at first, of any who had been through the course themselves. Each demonstrator himself does the two or three experiments which he will supervise during the year and becomes expert in the physics of each and its instrumentation. The folder on each of his experiments which he is given may contain fifty pages of manufacturers' information, directions on how to get best results within the time limit (for example, a routine enables a student to find an oven already heated to the correct temperature so that he can insert a Franck-Hertz tube into it, minimising both delay and deterioration of the tube) and directions as to how far the students should go with the experiment. And, at our visits, demonstrators said that they felt supported by the staff in the sense of being able to ask for help and advice, and not being left to get on with the job alone. The demonstrators' training had been restricted to the attainment of mastery of the subject matter and did not extend to teaching skills. However, apart from the experiment on errors and probability, we saw no demonstrator speak to more than three students at one time, and in this experiment the demonstrator was helped by slide-tape presentations.

Honours physics students at Bristol attend a laboratory with a
traditional circus of experiments in their first year, a similar
but more advanced laboratory in their second, and in their
third year they work on a single practical project.   We spent
two days in the Stage II laboratory, where students, working
one and a half days a week, typically complete seven out of
more than twenty available experiments.   No single feature is
very remarkable, but taken as a whole the laboratory has
distinct character.

Students work in pairs.   With a demonstrator, who may be
a staff member or a research student, a pair chooses an
experiment for which a set of apparatus is vacant.   They read
the script, look up references from it, use the apparatus, keep
a single laboratory note book between them, and when they have
finished they are interviewed by any demonstrator free to see
them.   He assesses their note book and their verbal presentation and defence of it.   Three times during the year each
student writes a detailed account of one of his experiments.

Most experiments are done in two main rooms, one of which
has technicians' rooms and dark rooms opening from it: most
demonstrators are usually in this room and the assessing interviews are there.   There are also two small rooms used mainly
for electronics experiments, which, in several important
respects, are managed differently from the rest.   The library
is near, and is often used by students in their laboratory time.

The scripts are short; often less than two sides of open-
spaced typing.   They state the object of each experiment in
general terms and direct students to consider alternative ways
of using the apparatus, about which there are sometimes
warnings and instructions.   Theory is seldom given; four or
five references are usually given to suitable books and papers.
Apparatus is modest in amount and complexity.   Experiments
are often designed so that accuracy is limited by something
which students may not alter, and an important part of the
students' task is to plan a good way round the problem.   One
script states: 'The present current balance has been devised
to illustrate some of the difficulties encountered in ...', as if
that were its only purpose.   Scripts offer choices, of which
students are expected to see the implications, for example in
an experiment with a coarse diffraction grating giving small
deviations: '... considering carefully what angle of incidence
will give you best results'.   Apparatus which students request
for supplementary purposes, such as a long tube so that gas in
a latent heat experiment may reach a known temperature before

its volume is measured, would be provided, but not left to inform later students of a problem.

In principle, demonstrators are available, and students, who are not sure what to do, can ask. A student who had done so, said that the demonstrator 'sort of vaguely pointed the way but didn't know much about it either, so we tend to ask other people' Similar experiences must be common because the character of the laboratory is not to have simple clear-cut answers, and teaching not to expect them is one of its lessons. Two students who had asked the director of the laboratory such a question said they had been left to find out for themselves, and that they got more out of it when they did. The demonstrator who 'didn't know much about it' may have partly pretended ignorance.

'We tend to ask other people' indicates a common source of guidance with fairly sharp conventions attached. 'We asked the groups who had done it before if ... there were any major objections to (a modification). So they said 'No, not really', and in fact one of them had actually done it... We didn't get that from them.' Major help, if we understood the conventions, should not be sought except as to matters which could not reasonably be foreseen; questions should ideally be framed so as to convey a polite interest in the questionee's success no less than concern for the questioner's; and a short answer may be given to anyone who asks grossly more than circumstances justify.

5.4.1 ASSESSMENT INTERVIEWS

The most important demonstrator-student interaction is at the interviews which close each experiment. These may last an hour or more. Describing one, a student said that the demonstrator (a research student) had asked questions '... on procedure and theory, what corrections we'd made, why we'd made them, what justification we had for doing it the way we did it, had we thought of doing it in a different way, what account we made of our statistics, and things like that ... A quite penetrating discussion.' His partner put in: 'They're trying to find out how much you learn...' and the first went on: 'Yes ... if you've understood it properly. It's also very useful as well, because if there's anything you aren't quite clear on... the demonstrator usually manages to find something you aren't clear on and quiz you about it... so that by the end of the... interrogation, if you like, you do end up getting a little bit more understanding about it'. A research student demonstrator, asked what he would count as bad, replied: 'If they haven't really understood'.

5.4.2   LABORATORY POLICY

The laboratory has reached its present state by a series of
modifications.   Ten years ago there were about three times as
many experiments, for each of which there was only one set of
apparatus.   But, whereas each student benefited only slightly
from the experiments which he did not perform himself*,
substantial benefit was expected from using only the best third
of the experiments, for which the apparatus could be trebled,
and for which the desired criteria could be more generously
met.   The policy with which the surviving experiments were to
conform may be summarised: to impart understanding of good
experimental procedure so as to enable students to plan the
extraction of maximum information in a given field within a set
of constraints.   Objects such as the acquisition of techniques
and the illustration of theory, while valued, were thought to be
less important, and not, in the second year at the university,
to be sought to the prejudice of the first.

   This policy had been consistently followed in the simple
apparatus, brief scripts, leaving students alone at their benches,
and the assessment interviews.   The special interest of this
laboratory is in the degree to which logic had been followed in
seeking a long term objective in the face of many necessarily
contrary short term ones.   For a demonstrator cannot say:
'Go and understand good experimental procedure', he must say
'Go and measure the viscosity of oil'   (or some such thing), and
put obstacles in the way of measuring the viscosity so as to
ensure that good experimental procedure shall be more fully
understood.   For having placed the obstacle, the demonstrator
is seen by the student to some extent as an opponent. Students'
comprehension of the demonstrator's dual character is among
the successes of the Bristol laboratory; all acknowledge what is
stated in the Stage II Physics Laboratory Notes for students:
'Good technique may be learned by overcoming the 'inadequacy'
of some piece of apparatus, and our experiments have usually
been designed this way, not out of pure malice, but because
almost always at the research level some piece of the apparatus
is working at its limit (it could hardly be research if it were
not) that is, it is 'inadequate' for the job.'

----
*It is hard to estimate what a student might gain from seeing
other students use apparatus which he has not time to use himself.   But in a laboratory such as the Clarendon at Oxford the
gain could hardly be negligible.

The inadequacy succeeds. Students criticise devices, and methods are crude enough to be criticised. They wish to improvise and improve. These wishes, like the wish to work unhelped, broadly characterise the maturity of second year students in relation to the insecurity of first.

But at another level, maturity bring scepticism. And, indeed, the student who, having watched the current balance swaying in the draught all day, never asks himself whether its designer is having a quiet chuckle at his expense, may lack some quality which he could need in other circumstances. But the actual costs of inadequacy, though less spectacular, are probably greater than those of adequacy, partly because that may often be bought off the shelf. Moreover, students stand in awe of the new and expensive and are fearful lest inadvertently they do it damage. Whereas mere simplicity, elegance, and good design are not so easily recognisable in apparatus as the fact that it was made many years ago or without regard to finished appearance. Students do not take such good care of what they do not appreciate and will risk damaging it in ignorance when otherwise they would have asked advice. It seemed to us that the Stage II students were thoughtless in this way, and that the technicians needed to be more active and vigilant in consequence. Their contribution was of a high order, and only partly recognised by the students as the expensive luxury it is.

5.4.3   ELECTRONICS

A more recent development has been the introduction of electronics which is seen as increasingly necessary for students, into a circus designed to encourage independence and scope for planning. It encountered some difficulties. If a student at a classical physics experiment consults a demonstrator because his apparatus has a serious limitation which he explains, the demonstrator may often be wise to congratulate him on assessing the problem so well, and to tell him that now all he needs to do is to go away and solve it. In a second year lab, a little of this generates a lot of independence, part of which consists of the student thinking the demonstrator knows no more than himself, with an acceptable penalty in maltreated or broken apparatus. But the same student may be only a child in electronics, often needing to ask for simple advice, and willing (or even anxious) to be watched at every stage. Electronics is also a problem for demonstrators, who cannot as easily give the more specific and more detailed advice which is wanted. At Bristol the quality of this problem was seen, but its extent could only be gauged by experience. Early electronics experiments, undistinguished from the rest of the circus, were less effective. Independence and initiative could sometimes seem more like vandalism, as when students, finding reverse insertion of circuits to power rails carefully protected against, just thought them too stiff and pushed harder, breaking the protection and reversing the supply. By the time of our visits, segregation of the experiments in small rooms with the constant presence of a specialist staff demonstrator had established a pattern in which frequent small increments of knowledge predominated, in electronics only, to everyone's satisfaction.

Yet the treatment of the electronics differentiated it fundamentally from the rest of the laboratory. Outside electronics, preparation for research was the purpose which both demonstrators and students repeatedly emphasised as being overriding, and the purpose for which the mere imparting of knowledge was to be sacrificed to inflicting need for decision-making. Electronics as a tool had been rated important enough for this priority to be reversed.

The latest phase in the laboratory's development is the introduction of some experiments on more modern aspects of physics without altering the general framework. But this was still at the stage of prototype apparatus being made.

5.5   SECOND YEAR AT EAST ANGLIA

The choices available to students at the University of East Anglia result in a small group, such as fifteen, attending the second year practical class, the number including both physics specialists and those combining physics with other subjects. The laboratory contains fifteen of what the staff sometimes call experiments and sometimes call projects. The students' main introduction to each experiment is its script, typically five pages including several references to standard books, which is available to be taken away at any time. What is done in an experiment is a matter for individual discussion between student (the students work singly) and demonstrator, for most experiments permit options and extensions. Each experiment has a fixed time; six specified three-hour sessions for a specialist and four of those sessions for a combined student. Each student does five of the experiments in two terms.

The director of the laboratory is assisted by only one other staff member. One of them is responsible for each session but the other often comes in. At each session the staff member is assisted by a post-graduate demonstrator, who attends every session for some months and is then replaced by another.

Apparatus is generously provided for each experiment, so that the student has to decide what pieces of it to use, and how to use them. For example, on one bench we saw a tube of oil, signal generator, low voltage supply, another oscillator, double beam oscilloscope, closed black box, AVO minor, travelling microscope, and a box with an integrated circuit in it. At another we noted this conversation between a student and a post-graduate demonstrator:

D You're still trying to find what that piece of apparatus is, aren't you?

S How can I use it if I don't know what it is?

D If you think of something to do you'll need to look around for suitable equipment. Anyway you can always look on the lid.

S Yes, it says it's a calibrated viscometer, but that still doesn't tell me what it is.

Then the demonstrator explained it. To another student we heard him say: 'How are you getting on? Have you worked out why you've got two meters yet? No? It was an innovation of your predecessor's.'

These notes were made at the first session of the fourth experiment. The remarks typify the approach of the demonstrator and also of the technician, who mixes freely with the students, and whose style is similar. At later sessions the problems were different.

In conversation, the director of the laboratory said to us: '... the basis of the lab is to try and make students independent and if that causes frustration, boredom or misery, well, that's how it is. If, at the end, they say they don't need me, then I've done my job.' Much of the apparatus had been deliberately chosen as being new to students and fairly complicated. We overheard, from the post-graduate demonstrator:

D You know how the phase sensitive detector works, don't you?

S No. I've read it (the maker's instructions) but it doesn't seem very clear.

D Well, the best time to read it is when you are going to use it.

Students are expected to make what they can of maker's instructions, but they are given help if they are found not to

be making progress. The technician was often asked for help
but he showed disapproval of lack of initiative, as when a
student, asking for tables for a thermocouple, specified it as
'the one I've got':

T   Don't you know what kind you've got?
S   No.
T   It's chromel-alumel

5.5.1   STUDENTS' ACTIVITY IN THE LABORATORY

Students were said to be encouraged to talk to and advise one
another. At the first session of the experiment, when there
were more problems than usual, we saw students wait in the
hope of getting help from the post-graduate demonstrator when
he was already engaged, and then go to other students instead.
At the later session students went to other students for advice
at times when neither demonstrator was in sight. Although only
three previous students would have done each experiment, there
was no hesitation about which student to ask in each case. The
consultations were short, entirely businesslike, and without any
element of relaxation from the task in hand.

All the students were present for almost the whole of both
sessions we visited. Not much of their time was used on anything but handling apparatus. For example, out of eleven
students easily visible at a particular moment, what each was
doing was: looking into a Fabry-Perot interferometer, placing
a lens for an image, listening to a demonstrator, writing in a
note book, adjusting an oscillator, switching an x-y plotter,
reading a meter, listening to a demonstrator, screwing a calori-
meter, turning a knob, adjusting an electromagnet. These
activities are typical and imply a fair amount of time spent outside the sessions on reading and other paper work, and the time
in the laboratory was clearly valued for doing the experiments.
Another indication of this was that one or two students were less
ready than in most laboratories to leave their work for ten
minutes or so in order to tell us about it.

Except their first experiment of the year, to which they are
assigned arbitrarily, students choose each new experiment
during the last session of the previous one. The six three-hour
sessions for the experiment are arranged within a two-week
period, which is followed by a two-week period when the
laboratory is closed. This is to shorten the intervals between
the sessions of a single experiment, making it a more intense
experience, and to lengthen the interval which is available for

writing up the last experiment and reading about the next. The 'good students' are accustomed carefully to exercise their choice of experiment, booking as soon as possible the experiment they wish to do, and they use the two-week interval to find out all they can about it. The 'weak ones' commonly accept whatever experiments are left over at the end without trying to influence what befalls them, and find out less in the interval.

5.5.2 THE HANDING BACK OF A NOTE BOOK

The assessing of students' work is done only by the two members of staff. A student hands his note book to whichever demonstrator normally deals with the experiment he has just done, probably during the two weeks when the laboratory is closed. The demonstrator may then read the book and he returns it at the next session, asking the student about details before making his assessment, and perhaps coming into the laboratory specially to do so. During one of our visits a student who had just finished an interferometer experiment was questioned. Although note books are intended only as a laboratory record - one experiment is chosen each year as an exercise in the art of communicating results to the scientific public - this student had left ten pages before entering measurements and had later put in an explanation of the theory in the space. He was questioned about why he had done this, how he had decided the amount, what book he had used. There were photographs of fringes, each in its envelope:

D   Which one shall I take?
S   It's up to you

He was confident of a fair degree of approval. The photographs had been measured with a densitometer, so demonstrator and student looked at a trace:

D   So we're on the one at 336 mm. How did you measure that?
S   With a ruler
D   A wooden ruler?
S   A plastic ruler, so I could see through it
D   You weren't tempted to use a travelling microscope?
S   No
D   Was that because you were lazy?
S   No

He knew the demonstrator knew he could justify how he had done it. A little later he conceded having believed the script on a minor point and was warned not to do so again. He admitted he thought the experiment was the most interesting of the three he had done but refused to give his opinion as to whether he had done it better than either of the others. The talk lasted more than half an hour. After it the demonstrator said that the student was a very good one whom it was rewarding to teach. And later the student said of the demonstrator: 'I respect him even though I argue with him.' Of the laboratory he said: 'If there's anything you want, you can get it. Last time I got a pulse height analyser worth about £2000... I was reading the script and I asked Dr X something about it, and he said a pulse height analyser would do the job and that he would lend me one.' Dr X was his tutor.

That Dr X should have been his tutor characterises the East Anglia laboratory. Everyone knows everyone else well. The student who has not yet comprehended the viscometer is not an anonymous student probably slipping behind, he is an individual of whom such things are known as the name of his girl friend, that he nearly chose to take the theoretical course unit, and that he dropped a glass thermometer last Wednesday. Demonstrators are not 'they' but 'he'. If anything is not right, its wrongness is not the manifestation of a system, it is the deficiency of a person. And when things are right, some known person has made them so.

We could not tell whether these intimate circumstances resulted in more or less or different experience of physics being acquired. They looked as if they were a poor training for overcoming the inertias of a big institution. But many of the great physicists of the past probably learned in conditions which, from the viewpoint of personal relationships, were much the same.

5.6   SECOND YEAR AT MANCHESTER UNIVERSITY

The second year teaching laboratory at Manchester has a familiar pattern. Students spend one day (effectively about six hours) a week in the laboratory and they are expected to complete seven experiments out of a circus of about thirty in somewhat more than two terms. Pairs of students choose an experiment for which there is free apparatus, read a script, augment their knowledge of theory from books, use the apparatus, work out results, and take their results to a demonstrator who interviews and assesses them, this process qualifying them to choose and start another experiment.

The building is recent, and is constructed with many small

secluded spaces so that one sees few people at a time. A Manchester student, who, after the first three weeks of the year, compared his laboratory work with a student from East Anglia, would feel that Manchester spared its students some slight harshness in the way of being impersonal. He would have a partner, whereas the East Anglian would have been alone. The Manchester pair would have worked in a space with perhaps two other pairs, whom they would have consulted about various things, not all to do with physics, whereas at East Anglia a larger number of people working singly might have discouraged that. Each day they would have seen the same demonstrators (probably two: a member of staff and a research student) whereas the East Anglia student would have seen different staff demonstrators on different days and would more certainly have needed to encounter the technician as well as the post-graduate demonstrator.

The point of these comparisons is that, with no other information, they might deceive one into supposing that Manchester was as small or smaller a department, whereas in fact it is an order of magnitude larger (about 130 per year). The whole second year laboratory is divided into five labs, each of which has its man in charge of it, in each of which a student must do at least one experiment, all of which are open at the same hours, and some of which are in rooms communicating directly with others. The labs are nuclear, electrical measurements, general, optics, and vacuum. The second year lab tutor co-ordinates them all, and he also demonstrates in, without being in charge of, one of the five labs.

So the pair of students, who have done their first experiment in a small room with two other pairs, who have been helped by the staff demonstrator who has looked after these same few experiments for several years, and who have perhaps decided that the post graduate-demonstrator had never seen their experiment before they had themselves, get their work assessed, are given a card on which that fact is recorded, and set off for another lab in which they know no one. Five times altogether they would go to a new lab. During the year they would go to a lab where they knew the demonstrators only twice; the odds are against their knowing the name of any other student working in the room they enter, although, once in the room, circumstances certainly favour acquaintanceships being formed.

5.6.1 COMING TO A NEW LAB

Our visits were made after Christmas when attitudes would have

become established, and enquiries, by students among friends
as to what experiments in other labs are like, could be helpful.
We overheard a demonstrator offer a choice of two:

S   Which is the quickest to get done?

D   I would say that the a. c. bridge teaches you more.

The demonstrator's studied evasion indicates that the question
should have been addressed elsewhere. But sometimes homework is done properly:

S   ... really, you've just got a list of experiments. The
available experiments, say there were four or five to
do and its a matter of looking at the names and deciding
what you'd like best by the sound of it, and so on ...

I   ... with nothing much to judge on... or have you got
other knowledge about them?

S   You tend to find out from other people what they think
of certain experiments, and I think, you know, that's
a good source of whether that's a good experiment to do
or not

I   Had you heard this was a good one?

S   We'd heard it was OK to do. This particular one
involved a lot of maths because the experiment itself
was quite simply straightforward, but ... it was the
sort of experiment where you could do a lot of work
on it if you wanted to. It wasn't a closed thing. You
could take it further. So we thought it was a good one
to do.

5.6.2 SCRIPTS

Each of the five labs has its peculiarities and an 'experiment'
in one lab may differ from an 'experiment' in another in many
other ways than the piece of physics it is about. One lab
(nuclear) for reasons probably including the equipment's
complexity, has no scripts. Two students who had completed
a nuclear experiment were asked:

I   How did you find out how to do it?

S   Well, the demonstrators in that were particularly helpful
... sometimes they'll simply refer you to text books ...
you have to go away and sort of try and pluck out what
you need from it, but they were very helpful and
explained what we were trying to do... always useful
before you start to do an experiment.

I   Was there a sort of script to it?

S   Not in that case.   The demonstrator... he wrote as he explained and we kept that.

Scripts in other labs varied widely in length and in the sorts of information they gave.   One, on the field effect transistor, had eight pages, three references, optional extras, and advice about real (as opposed to ideal) properties of components likely to be used.   Other scripts were less than a page.   We encountered two students taking advice from another as to how and whether to do a particular experiment.   The advisor recommended it, saying that it was interesting, and that it 'ties in a lot of concepts', and he explained in a general way about the apparatus and in particular about one piece that had a fault. Discussion lasted five minutes or more, and was informative, unhurried, and uninhibited.   The three page script was half algebra and gave no references.   The two students decided to do the experiment, and negotiated a post graduate-demonstrator to tell then the official story.   He came.   His words, as he gestured towards the bench, were:

D   That's the equipment there.   One of the boxes is a bit dodgy.

and this formally marked the start of the experiment.   It was not clear whether the students expected or wanted more.

The two students in the nuclear lab, immediately after the dialogue earlier on this page, were asked:

I   ... did you get much help during that time from the demonstrators?
S   Yes, a sight more than some of the other experiments we've done.
I   You seem to approve of that... or do you not?
S   It got a little difficult at times... the demonstrator would come along and say 'How's it going?' and you... it got on our nerves after a while... you'd know he'd come in every half hour.

These students much preferred being left entirely alone once they thought they knew what to do.   It is very likely that the post-graduate demonstrator whose perfunctory instruction is noted above had been rebuffed on some occasions when he had tried to do more.

5.6.3 EXPERIMENTS

Students find out or fail to find out about their experiments in
a variety of ways. Correspondingly the experiments themselves
differ. Recently designed experiments tend to teach interesting
pieces of physics. Those experiments which appear to have
survived from the earliest period are simpler and more austere
in their hardware, and more sternly algebraic in their scripts.
Their long survival may denote their consonance with the
laboratory's aims, which were expressed to us as being
primarily to teach experimental physics in the sense of how to
do experiments rather than to back up lectures, valuable as that
is. More concisely, it was aimed that students should be
brought to appreciate the best ways of using equipment - 'method
rather than techniques'.

Students more often spoke of interest or otherwise in the
physics itself. Two, who contrasted the second year laboratory's
concern about physics with the first year laboratory's concern
about errors, had chosen the experiment on the field effect
transistor in the general lab because they liked the sound of it.
They said:

S 1 ... it was more plotting characteristics... it wasn't all
that interesting till the last stage, where we built an
amplifier and looked at the frequency trace.
S 2 It could have been interesting if we'd been another week
or so on it. There's a lot of things you can do with
that which would have been worth going on with. We
spent too much time on the d. c. characteristics. There
just wasn't enough time to do it.

This regret that more time could not be spent on experiments
was common, but whereas it can be taken as a fairly reliable
expression of interest in the physics to be learned, its literal
meaning may need to be considered in conjunction with the
pressure of non-academic calls upon a student's time.

Two other students, whose adverse reaction to their experiment is untypical of the lab in which they did it, but typical
except in its articulateness of reaction to experiments in general,
described what they had done:

S 1 ... there's a neutron source in the middle of a tank.
The tank's full of water so neutrons get attenuated by
the water and then by putting indium foils in a tube
through the tank you can find the variation of the
neutron flux as a function of the radius. So, that's
basically what you do. Well, you find the half-lives

and... this particular experiment I think probably the worst experiment I've done. Because it's just boring. You know, it's just a matter of taking readings and waiting and things like that, because you have to leave the thing in there for an hour for it to...

S 2 ... there's several experiments on this tank, and just as a comparison, the experiment we're doing: we put those foils inside the tank in a certain position, take them out after a measured length of time when they've had a certain time to irradiate and then measure their activity in a chamber with a G.M. tube. The group next door, though, would be putting a G.M. counter straight into the tank, so they don't go through all this process of having to irradiate the foils and let them drop down to a very low activity before they can start irradiating them again. So you might say already they are using a far more efficient method of doing the same thing than we were, and it's quicker... a bit more interesting, I suppose.

S 1 And probably more accurate as well.

Arguably these students display, by their dissatisfaction, more profit from the laboratory in terms of its own professed aims, than they would have done by uncritical approval of the experiment.

5.6.4 INTERVIEWS

The interviews which close each experiment are an important, and in some labs the main, contact between demonstrator and student. Clearly the more poignant impressions made on students' minds were often made by them. Two students, speaking about interviews generally, said:

S 1 ... and he'd also throw in a few questions... to see if we've read up anything more, any background. Sometimes those questions can... they completely floor me... they're beyond what I ever thought the experiment might have been about. But usually they are... interviews are interesting but... I find them frightening... because it's mostly on how the interview goes... that will affect the mark you get at the end of the experiment.

I If you do it well or badly, does it make a big difference to the mark you get?

S 2 Yes.

S 1 If the demonstrator thinks all you've done is to repeat what's been on the script, and not thought about anything else, then he'll mark you down.

S 2 And tell you about it in sarcastic terms.

S 1 Yes.

S 2 And there's no misunderstanding.

I Yes... but is it the sarcasm or the marking down that matters more?

S 1 I don't know quite how important the marks are, actually. Don't even know where they go to.

S 2 They affect your final... er... mark... I think it's the sarcasm more than anything. You feel as if you ... with some demonstrators anyway... you feel as if you've spent three or four weeks slaving away at experiments and he just shoots you down destructively.

This description does not support the view of students as being largely motivated by assessment procedures and carefully calculating how best to play the system. Rather it supports the view that interviews' main results are subjective - that the sudden realisation, in a face-to-face confrontation, of one's own defencelessness, is much more to be avoided than only getting five out of ten. And conversely, as the second student went on to say of one particular interview, 'It wasn't just that it went well. I think I learnt more out of it.'

Another pair, talking about note books and the assessment interviews, were asked:

I ... do you reckon it's the book that's being marked by the demonstrator, or is it you that's being marked?

S 3 It's us. Well, he is going to be impressed if you've obviously looked up a lot of things and written it down in the book, but I think generally in interviews he's more concerned with what you did in the experiment, and it's generally questions about the experiment and not really related to things you've got in your books.

S 4 But this again brings us to the point about what you reckon the demonstrators should be after, and I think that in general the people in charge (staff members), normally when they give us an interview afterwards, they tend to be after the right things in so far as they're after your understanding of the experiments and your results.

The student went on to compare post-graduates with staff demonstrators, agreeing with the judgment of a student in another lab: '... the latter are better. They look at the way you've treated it and at your understanding.' The staff demonstrators' superior skill is partly acquired by constant practice - the commitment of 150 hours in a year to demonstrating is a large one and no doubt engenders economical techniques and professional attitudes.

Whether choosing an experiment, doing it, and perhaps even when getting it marked, the person whose importance is greatest to each student is his partner. By comparison with the partnership, other personal relationships formed in the laboratory are ephemeral. Students spoke of 'us' with trust and certainty, but of 'them' as of objects to be observed and evaluated. The partnership, and the constant discussion within the partnership, with its need to find terms in which the physics and the people can be discussed, must be the most valuable secondary feature of the laboratory.

6. Progress reports 1

6.1 INTRODUCTION

This chapter is built around accounts of four contrasting laboratory courses, written by the teachers responsible for them, with a commentary relating their ideas to those else-where in the book.

The first two accounts concern first year work, the first being about the design of a year's course progressing from programmed experiments to small projects. The second, a special case, concerns a 'course' lasting just one afternoon, operating therefore under very tight constraints.

The second two accounts both concern 'unit' laboratories, a concept described in greater detail in chapter 8. Each is a laboratory course concentrating on one topic or theme for a few weeks. Both are for second year students.

No real life laboratory can operate without adjusting to and compromising with a variety of practical limitations and pressures. In the last chapter, some of the ways in which laboratories can evolve within - even exploit - these boundary conditions were looked at. In the present chapter, all the accounts are of laboratories in which some decisive, definite change was made; in which the design of the whole laboratory was in some sense remodelled. This difference has several consequences. A need to think more explicitly about the purposes of the laboratory was often felt, and the accounts reflect this. New and unexpected consequences tended to arise more often than they usually do in laboratories which change in an evolutionary way.

The purpose of the chapter is not to propose a debate between revolution and evolution. All the laboratories described here changed and evolved too, and were decisive in the way they changed more for local reasons than for any belief in suddenness of change. Rather, its purpose is to expose in another way, as before by example, some of the many decisions and ideas that lie behind the structure of any teaching laboratory.

6.2   STARTING FROM SQUARE ONE AT BATH

Patrick Squire writes:

> It is not often that the opportunity arises to redesign a
> laboratory course from scratch.   Just such an opportunity
> did arise in the Bath first year laboratory in 1970.   The
> pattern of laboratory teaching in operation until then had
> been based on five or six groups of traditional experiments,
> each group under the supervision of a subject expert, free
> to organise the activities within his area.   Looking back,
> one can see some resemblance to the unit laboratory system
> (see 6.4 and 6.5).   However, the virtues of that system and
> the advantages it can bring to a large department were not
> realised in the context of a student intake of 30.   The over-
> riding impression was of inefficient use of staff time, a
> heavy burden of report writing, and confusion about aims
> among both staff and students.

> In order to improve this situation, two members of staff
> devised a completely new course, which was approved and
> developed over the summer of 1970, ready for launching in
> October.   The guiding principles on which the design was
> based were:

> > The traditional experiment either has no clearly stated
> > aims or attempts to fulfil too many at once.

> > Two important aims of laboratory work are sufficiently
> > distinct to justify their isolation as separate activities.
> > These are the acquisition of techniques, and illustrating
> > important theoretical concepts or physical behaviour.

> > Some project work is highly desirable for the synthesis
> > of skills developed separately, for motivation, and for
> > the insight given into real-life physics.   However, much
> > frustration can result unless the student is suitably
> > prepared, and the project is carefully chosen.

> In order to implement these principles, four types of
> activity have been used:

> > Quasi-programmed instruction in techniques (T)

> > Quasi-programmed demonstrations   (D)

> > Supporting lectures and problems classes   (L)

The course structure is summarised in the following profile:

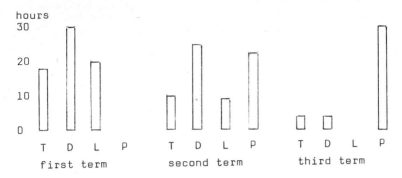

COURSE PROFILE: TIMES SPENT ON ACTIVITIES

The exact proportion of time allocated to the four types of activity is obviously negotiable. For guidance, the times shown allow for about eight techniques, ten demonstrations and three short projects. The choice of topics to be included was originally made in consultation with colleagues.

The quasi-programmed work uses small duplicated booklets of instructions and questions, which take the student step by step through an experiment. An example of the techniques taught is the use of an oscilloscope. An example of ideas illustrated by a demonstration, which again the student does himself step by step, is the use of resistive paper to plot equipotentials and field lines (Squire 1974).

Written instructions were drafted in the early summer and tested on a group of local sixth-formers, who were very willing to co-operate, and provided a realistic sample of the type of student starting the course.

The use of carefully prepared written instructions achieves two useful results. It relieves staff from repeating the same routine instruction to many students. It also helps to ensure the effective acquisition of techniques and successful demonstration of phenomena. Originality and initiative by the student are not required, but they may have questions to answer which require considerable thought and understanding. Log books are kept, and inspected at the end of the year, but reports are not required for the programmed activities.

The supporting lecture and problem classes contain a fairly conventional assortment of topics, such as data analysis, graph plotting, report writing and design of experiments. There is a strong emphasis on numerical problem solving, using calculators.

Thus, by early in the second term students are ready to embark on the first of their three projects. Each project is designed to be capable of yielding useful results after about eighteen hours of laboratory time. Topics are chosen by the students from a list, but they are encouraged to work in more than one subject area and at more than one type of project (for example, hypothesis testing, apparatus design). Students normally work in pairs, but with a different partner for each project. A full report is required for each, and these constitute the major part of the year's assessment in the laboratory. It is the overwhelming opinion of the staff involved, and of the students, that these projects are a worthwhile and enjoyable activity. However, they require careful organisation and the full co-operation of the technicians if they are to succeed.

One advantage of this course structure is the relative ease with which changes may evolve within it. The main factors leading to change have been : developments in technology (for example the introduction of digital measuring techniques); changes in overall course structure (for example the intro- duction of a Physics with Geophysics degree led to the inclusion of projects relevant to it); suggestions by staff involved in running the course; and results of a concluding questionnaire issued each year to all students. This last has been a most useful source of feedback, and has resulted both in the deletion or improvement of existing material, and the addition of new material.

The staffing ratio has been at a level of one or two lecturers plus one postgraduate for classes of between 22 and 43. For the experimental work, two periods of three hours per week for 25 weeks are used. In addition, two hours for 15 weeks are used for the supporting lecture/problems course. Three technicians are available. The laboratory space used consists of two large areas, each approximately 140 square metres in area, and sixteen cubicles, each approximately four square metres. These same areas and much of the apparatus are used at other times by other classes.

In summary, the Bath first-year laboratory course provides

within a fairly rigid framework a range of activities, varying from tightly structured technique instruction to project
work allowing considerable freedom. It has run for six
years and has proved to be adaptable and effective. The
advantages of this type of course are most likely to be
realised with a class of 20 - 40 honours degree physics
students.

6.2.1 COMMENTARY

One obvious feature of the Bath development is its reliance on
an analysis of the problem. Aims have been broken down into
more definite parts, and the structure of the work reflects this
analysis. A good deal thus hangs on whether it is a good
analysis.

As do many others, they distinguish relatively definite, often
short term objectives from longer term, less analysable ones.
The first are dealt with by particular exercises, each intended
to achieve only one or two definite things. The second are
dealt with by a project style of work.

The second feature is the belief in progression; in changing
the style of work through the year, as students learn more, and
perhaps change their attitudes.

Thirdly, the laboratory is seen as a whole, with contributing
parts in a total design, not as a collection of equivalent
components.

Some visits were made to the Bath laboratory at the time at
which most of the work was in the form of quasi-programmed
activities, and observations made at those times may cast some
light on how these work out in practice.

It might be reasonable to think that providing more or less
step by step instructions would lead to more (and more systematic) handling of apparatus, to less need of assistance from
demonstrators and more reliance on instructions, and to more
definite ideas on both sides as to what might have gone wrong.

On the whole these expectations are supported by observation.
Out of observations of what students were doing at five minute
intervals over an hour and a half, on no occasion was a student
sitting in silent contemplation, or doing something evidently
irrelevant. On only a few out of over a hundred observations
was a student talking with a demonstrator. The majority of the

activities were talking, taking readings, reading, and writing. Students differed markedly from one another in the extent to which they talked to others.

A portrait of what a student did over three hours with a programmed techniques demonstration (the oscilloscope) may be worth painting. The student arrived promptly, sat down in front of apparatus and booklet, opened the booklet and began reading. For the first ten minutes, he read an instruction from the booklet, performed it, and read another; a typical cycle taking half a minute, and involving for example finding a gain control and setting it at a prescribed value. When the next-door student asked how to find one of these controls, he was able to tell him. The student observed also asked the other questions of the same kind. There were about half a dozen such exchanges in the three hours, more of them being near the beginning.

The student wrote a good deal in his notebook as he went along, typically filling a page or more after a half hour sequence of working through the programme, and using the programme page as he wrote in the notebook. Thus after nearly an hour, he had filled two pages of notes, and had made the oscilloscope produce a spot or a horizontal line.

There followed another long span of similar work, now with a signal generator connected to the oscilloscope. It was noticeable that the student first looked at the generator when he came to it in the programme, after nearly an hour at the bench.

During this phase (about an hour) one measurement was not what the programme said it should be. The student went and fetched the technician, who told him to record the value he really had, but also fetched the member of staff in charge, who went through ten minutes' worth of teaching about peak and rms values, and about how to check calibrations.

It was noticeable that the student did not repeat previous mistakes, of connections for example.

After two hours, he leafed through the remaining pages of the programme, and settled down to the last third, which concerned producing Lissajous figures, obtaining them and drawing them. After three hours he had completed the programme, and had filled almost six pages of his notebook. In this time he once asked for help on a specific point (see above), was once approached in passing and given a hint, and twice a demonstrator came by or asked if he needed help and went off without any interaction. The student left a few minutes after the scheduled time.

All this would seem to add up to reasonably solid, systematic work, in which the student successfully followed the path laid down. It did not, nor was it meant to, involve him in taking any significant initiatives. Apart from the considerable time spent writing, a substantial fraction of time was spent doing something with apparatus, and very little in thinking about what to do. The programme seemed to function, as one might expect, as a pace setter.

When students were interviewed about the lab, they tended to see it as something rather definite:

'It's a good idea to get a basic grounding which is quite firm'

and as not too pressurised:

'I think they try to let you in slowly'.

Some reacted, but not strongly, against the uniform programmed work done equally by all, whatever their previous knowledge:

'Perhaps it's too spoonfed - it's not necessary to put any thought into it. But I think it helps to bring everybody to the same level.'

At the same time they tended to think they had learned what they should have:

'We'd feel quite confident about how to do that now'

It was noticeable that a good number of students had a less sure grasp of the pattern of the whole course than is offered in the description given previously by one of its authors. The projects (then still in the future) were seen as rather hazy entities, and some had the impression that they would inevitably involve just those techniques that were currently being taught. One student said that the probable purpose of the projects was:

'to assess how we've learned the techniques and demonstrations...'

a view which reflects some, but by no means all, of the intentions behind the laboratory.

Rather tentatively, it might be proposed that the very definite form of instruction in this early stage of the lab gives students an impression, partly true and partly false, of learning a decisive set of individual things, each due later on to find its

115

specific application.

6.3 A SELF-SERVICE LABORATORY MINI-COURSE

The next account describes an unusual situation that, perforce, used sharply defined and streamlined experiments even more than the Bath laboratory.

Sid O'Connell writes:

> The laboratory filled an available single four-hour time slot forming part of a common introductory course taken by about 150 first year science and engineering students at Surrey. It was designed to be a highly organised efficient circus of four packaged and streamlined experiments, but in which the students also had a choice of those experiments which best served their needs or interests.
>
> Typically, in one four-hour session, a student might do a data-processing exercise (compulsory) and then choose to learn to use an oscilloscope, to measure the charge to mass ratio for electrons, and to investigate the properties of a potential barrier using a mechanical system.
>
> Two demonstrators normally ran the lab, with twelve to sixteen students per session. At a pinch, one demonstrator could have managed.
>
> The intention of the design of the laboratory was to provide, although in miniature, a course in experimenting rather than a collection of experiments. That is, the experiments were designed so that they all served one or more aims from the set of aims of the laboratory, and so that students could choose from the available experiments a selection which represented some balance of aims.
>
> On arrival, students were given a booklet which told them the purpose of the laboratory, emphasising that it was a place to learn things efficiently. The nature and aims of each experiment available were set out, and students were asked to choose three experiments which seemed to them either to cover a broad spectrum of aims or to concentrate on those particular aims most relevant to him or her. They made these choices on 'course design sheets' which were collected and used to work out a timetable for the session while students got on with the paper and pencil data processing exercise. In practice, the timetabling proved a simple

matter, and little duplication of apparatus was required.

The timetable had to be kept to rigidly, with students changing experiments after every hour. At the end of each, they filled in a brief feedback sheet rating its difficulty, interest, and relevance to their own courses. Information on any particular difficulties was also collected. Students were asked finally how well the experiment fulfilled the aims it was intended to in the mini-course, in part so as to focus students attention on the purpose of what he had just done. All this information was used systematically, year by year, to modify the experiments.

Each experiment was designed to relate to one or two - at most three - of the aims of the laboratory. Everything in the experiments outside these aims was minimised. Thus, for instance, if an aim was to train in observation, the experiment was contrived so that the phenomenon to be observed was certain to be seen, regardless of skill or effort. Again, for such an experiment no setting up of apparatus was required - for example, the spectrometer was provided fully adjusted and in focus. Only if the aims of an experiment included training in report writing was writing a report expected.

That is to say, the experiments were all highly streamlined. This was in part so that they clearly served very specific limited aims, and in part so that they could be completed in an hour each. For the latter reason, the scripts were programmed, with students working through the experiment step by step, normally filling data and observations into spaces provided in the script.

In running the laboratory, detailed evaluation was treated as part of the normal teaching routine. Where difficulties showed up, attempts were made to remedy them, with visible success. Overall, evaluation indicated that students, as intended, considered that their time was used efficiently.

6.3.1 COMMENTARY

It is necessary to recognise clearly that this laboratory is seen by its authors as part of education as a technology. They were not satisfied with an intuitive approach to its design, nor with a casual or subjective approach to evaluation and improvement. Rather, they selected and defined aims, having previously looked into differences in the ways staff and students saw and rated various aims (Boud 1973). They collected data about the experiments systematically, and used it to check on intended improvements (O'Connell, Penton, and Boud 1974).

A visitor to the laboratory might well be struck by this technological aspect. Everything has its purpose, and whatever does not serve that purpose is pared away. The experiments are as functional as some modern architecture.

There is some evidence (which the staff of the laboratory would be the first to admit) that students do not entirely share the staff's view of the close connection between aims and experiment. To a considerable extent, students choose experiments out of interest, and not for the worthwhileness of their aims.

The laboratory poses another problem worth thinking about: the problem of how far an example of something counts as a contribution to it. Thus the oscilloscope exercise in the Surrey laboratory is said to contribute to the two aims of teaching basic practical skills and of familiarisation with important apparatus and techniques. In this case, the idea of contributing one piece to something larger looks reasonable. The experiment in which a spectral line is identified is intended to contribute towards three other aims: training in observation; use of experimental data to solve specific problems; and to stimulate or maintain interest in the subject. It might be argued that it is much more open to doubt whether such aims can usefully be thought of as cumulatively built up part by part. Were the student to find this experiment very boring or very interesting, it is equally hard to be sure whether that would affect his interest in the subject in any very definite way. Similarly, the ability to see and identify a spectral line might not contribute at all to other matters which could equally plausibly be called skills of observation, such as noticing wave patterns on water, for instance. The point is raised here, not in criticism of the Surrey laboratory, but to point to a matter of some importance in thinking about the claims anyone might make for his laboratory.

The Surrey laboratory also brings out particularly clearly the trade-off every experiment makes between sharpness of purpose and freedom or opportunity. On the one hand, experiments so streamlined as those just described do not allow - nor were ever meant to - opportunities for the student to do something of his own or learn something unforeseen. On the other hand, many experiments in other laboratories, while allowing such liberty, can seem to students to have no clear purpose other than work to be got through, or can seem (and even be claimed) to serve an impossibly large and even conflicting range of aims.

6.4 A SECOND YEAR ELECTRICAL MEASUREMENTS UNIT LABORATORY

The unit laboratory system used at Birmingham is described more fully in chapter 8, where its organisational aspects are considered. Here an example is described. Briefly, a unit laboratory is one of a number of short laboratory courses, each occupying a few weeks, attended by students in groups. Thus in a large department, one might have four unit laboratories, each occupying a quarter of the students for a quarter of the year. Unlike a circus laboratory, attended all the time by all students, with equivalent experiments done in rotation and in no special order, the unit laboratory offers its designer the chance to plan a sequence of work, varying from week to week, building towards some end-point.

The laboratory described below is for second year students of physics in the University of Birmingham, and deals with electrical measurements. Robert Whitworth writes:

This laboratory class is held in an unnecessarily large room that once housed the main second year traditional circus. An area is set aside with a white board, tables, and chairs for demonstrations and discussions; the rest of the laboratory is furnished with normally empty tables, and with eight large cupboards which are open to students and contain the main stock of instruments and components.

Six groups of about twelve students normally attend the laboratory during the year. Each group works for eight hours a week for five weeks, and in any one week two groups use the laboratory on different days. The laboratory is run by three members of staff, who individually take complete responsibility for two of the six groups. At a later stage in the year these same staff take their share of running student projects using the same room and equipment.

At a fairly early stage in the design of the laboratory its aims were formulated in something like the following terms:

To provide students with further experience in the properties of d.c. and a.c. circuits.

To familiarise students with a range of both traditional and modern electrical instruments and the ways in which they can be arranged to form measuring systems. The instruments include in particular potentiometers, digital meters, recorders, oscilloscopes, amplifiers, filters, integrators and phase-sensitve detectors.

To provide practical experience in treating electrical devices as 'black boxes' with input and/or output characteristics.

(Most important) To give students some experience of what is involved in designing an experiment. This is to be done with particular reference to the selection of instruments and the design of circuits, and the design is to be seen through to the completion of a real experiment.

It was clear that students would need both to be taught about the instruments and to make use of them in designing experiments which would be seen to be real experiments in physics and not just exercises in electrical measurement. The time was therefore divided between an initial series of demonstrations and set exercises, followed by a major experiment in which the students were fully involved in the design of the system to be used. In order to leave adequate time for the second and most stimulating part of the work, the earlier part has been forced into a rather tightly scheduled set of tasks. This may not be entirely satisfactory, but is an example of the kind of choice which one is forced (though also free) to make in designing a unit laboratory. The work is organised in three phases, described below.

PHASE I LEARNING ABOUT INSTRUMENTS

In this nine-hour phase the demonstrator leads three demonstration sessions of 1 to $1\frac{1}{2}$ hours each in which he explains the kind of measurements that can be made and the conditions that an instrument must satisfy to be suitable for a given measurement. In the first two sessions he introduces the instruments available in the laboratory, including some such as amplifiers, electrometers, X-Y recorders and frequency counters, which students have probably not used before. The third session is entirely about noise and the use of a phase-sensitive detector for the recovery of a signal from noise.

In between these demonstration sessions there are two periods when the students do some very simple exercises like measuring the value of a high resistor using any of the instruments they choose. These provide opportunities for them to select and handle the apparatus, see its capabilities and discuss the instruments and their results.

PHASE 2 EXPERIENCE USING INSTRUMENTS

This phase is a circus of four experiments all done by every student, chosen to illustrate the use of instruments in ways particularly relevant to the design work of phase 3.

Fairly detailed instructions are provided, but they still require the student to make decisions about points of fundamental importance such as how to arrange the components in the arms of an a.c. bridge. The time allowed is fixed (one and a half or three hours) and some students are consequently rushed, but the aim is more to learn how the apparatus can be used than to complete detailed experiments.

PHASE 3   INSTRUMENTATION PROJECTS

The final twenty-two hour phase is occupied with experiments for which the basic mechanical apparatus is provided, but without any electrical instrumentation. They are chosen so that where possible they can be done in several different ways, but any attempt to tackle them soon reveals that there are stringent design criteria which bring out the principles of design. Examples are an experiment to measure the specific thermal capacity of an aluminium block suspended in vacuo with an electrical heater and a resistance thermometer mounted on it; an experiment to measure the susceptibility of a paramagnetic salt by observing the change in mutual inductance when it is placed inside a pair of coils; and an experiment to determine piezoelectric constants by detecting a change in capacitance of a capacitor one plate of which is mounted on a piezoelectric specimen. There are six such instrumentation projects for each group of twelve students.

In the first place the projects are treated as paper exercises, which students consider and then discuss in the class as a whole, aiming to estimate the size of the effects and to suggest techniques for measuring them. At this stage students often say that the experiments seem impossible. This does not worry us. The member of staff asks questions and draws attention to design methods. Even if the class do not themselves design the experiments, they have at least shared in doing so and seen some of the decisions that must be made.

Each instrumentation project is then given to a pair of students, and they have two weeks to get it working and to obtain results. In this time they often improve on the initial design. They also obtain experience in fault-finding and come across such practical problems as mains pick-up, earthing, drifting of amplifiers, stray capacitances and the like. The demonstrator is kept pretty busy. He tries to make sure that all students obtain some results. At the end many students are gratified to have achieved what they at first considered to be impossible.

IMPRESSIONS

Students find this laboratory hard work, but they seem to realise that it is important, relevant and usually interesting. They are in general prepared to put in a reasonable amount of hard work, and attendance has been exceptionally good. This is probably because there is something definite to be done in every session, and they feel they would lose something significant if they were absent.

The staff who set up the laboratory have found it a most stimulating experience. They would be sorry to go back to a traditional laboratory, not least because the instrumentation projects are constantly being done in new ways so that genuine new features or problems arise every time the laboratory is run. The group present at any one time is small enough for there to be good contact with every student.

Staff coming newly to the laboratory have also found it stimulating, but a crucial question about such a laboratory is the extent to which, as the originators leave, new staff will be able without unreasonable effort or expense to remould the laboratory so that they can identify themselves personally with the way it is run.

6.4.1 COMMENTARY

A notable feature of the electrical measurements laboratory is that it consists of a variety of activities planned to develop in a sequence towards a definite goal.[*] In terms introduced in chapter 4, it is at least somewhat singleminded (4.5), and its tasks also span the range from exercises at least to improvisations, and perhaps to projects (4.3). What is interesting is the way it uses many of the styles of work possible in the laboratory, weaving them into what is hoped to be a whole.

It is important to ask whether students see the aims of the laboratory and the value of the various activities as the staff do. As to the general structure, they seem to do so. A student typically outlined the five weeks as follows:

> 'We started off with a general discussion of equipment, and then did small experiments just to test it out. ... Then we did bigger experiments which take an assortment of equipment which has to be matched. Then this progresses on to the project.'

This would seem to be a rather accurate reflection of the staff view. At a deeper level, some students see rather clearly the kind of thinking the laboratory is trying to develop:

> '... thinking out different methods. I've realised quite a few things I wouldn't have thought of before. You've got to build it up gradually... You've got to keep on using things until you get used to thinking of everything - is it sensitive enough? high enough input impedance?'

All at least seem to get the general point of the lab, in some such terms as:

> 'You're learning about measuring things.'

The use of demonstrations to the group and of discussions with the group are features not so often found in other laboratories. It is probably true to say that it is in these that the staff feel least at ease, feeling themselves lacking a certain deftness of touch. Several students saw value in what they could get out of discussions without participating:

> 'I took quite a lot of notes... because (the discussion) amplified my ideas in a way I hadn't seen.'

> 'I appreciate going through a discussion like that even if I don't contribute - I'm a person who sits at the back and listens.'

The discussions concerned plans for experimental design. As staff said, and as students saw, their main outcome was often to emphasise the thinking that had not been done. One student said:

> '... nobody really knew what was expected... but through discussing it like that, and seeing just what we didn't do, then we'll be able to do it now.'

An assessment was required, and part of it was based on contributions to discussions. Several students were unhappy or ambivalent about this feature:

> 'Self-confidence is going to be the thing that governs a good or a bad mark, and that's not right.'

while others rather acutely saw what might be going on:

> 'I suspect that when you talk, for marking he gets an

idea of how with-it you are - if you know what you're doing - it should be obvious when you're speaking.'

The final instrumentation project was seen as valuable, and students' comments confirm Robert Whitworth's remark that at first they think them impossible, and are pleased by their subsequent successes. The work on the project bulks large enough in students' minds for them to distinguish sharply the value of the work itself, and the merits of the report written about it:

'One bloke got on his report, "This experiment was more successful than I had anticipated", but if you knew the time that had been spent cooking the results... he changed it so much that a lot of what he wrote wasn't what he did.'

But students see behind such matters too:

'You explain things you didn't realise at the time. You say I did so and so because of this, when you just did it by guesswork. Still, if you do that, and get it over, it means you come out at the end knowing what you should have done... You can't fiddle understanding... they'll soon see that.'

When students had finished their instrumentation projects, they seemed to be able to talk reasonably confidently about how their measuring systems worked and what the main problems were. It was difficult to find any who had no real idea what they were talking about. Further, for some at least, the instrumentation project built confidence about future larger scale project work:

'I'm looking forward to the long project. I wasn't until I did the project in the electrical measurements lab. I thought, a feasibility study: what on earth shall I do? And yet that project showed me how I'd go about things.'

So far as staff other than the originators of the laboratory are concerned, it must be said that Robert Whitworth is right to put his finger on the need to find ways in which they can make the design their own. The structure is tight-knit, and not obviously necessary or natural to those who did not participate in its design. They may approve, but still not feel sure of it, even though much care has been taken by having them sit through one or more whole five-week sessions before doing one themselves. So one hears:

'I wasn't in on the beginning when these things were planned so I don't know the actual reasons why they thought of doing things the way they do... I think it's a marvellous lab - wonderfully thought out... I don't think there are many improvements I could make.'

Finally, the electrical measurements unit laboratory raises a question of some general importance, in a rather direct way, because it has so carefully been planned and organised to meet students' needs in so many respects.

The point is that it, like many university laboratories, is a pre-planned, pre-arranged system. The needs to which it responds and for which it caters are necessarily anticipated needs, not ones discovered and dealt with on the spot to more than a minor extent. Of course, defects are discovered and are put right, so that the system will on the next occasion meet needs better.

That is to say, within this laboratory, no real possibility exists of finding out what it might be best for an individual student to do and suggesting that he does it. An adaptive system is defined out of existence: needs are supposed to be correctly anticipated, not to be discovered. For example, in the electrical measurements laboratory, the teacher is not required to decide when a discussion might be opportune; instead, the moment planned for it arrives.

These reflections, however, raise larger issues and will be taken up again later.

6.5 USING A UNIT LABORATORY FORMAT TO TEACH ELECTRONICS

The previous report (6.4) described a laboratory containing an organised sequence of varied activities, lasting for a few weeks, leading to a goal. Something very similar would serve to describe laboratory courses in electronics in a considerable number of institutions, and the next report describes one of them. Roy Davies, from Royal Holloway College, writes:

When demonstrating with our third year projects, helping students who were making use of electronic circuits, I repeatedly found that they had no real framework of know-ledge in this subject. To be sure, students did attend lectures on electronics in their first year, but it seemed that they did not relate the theory properly with practical problems.

In trying to do something about this I was anxious to avoid the problem of individual students performing sequences of relatively unconnected experiments being helped indiscriminately by a variety of demonstrators without any close personal supervision. It seemed to me that the unit laboratory format would be very suitable for a second year theory-cum-practical session in which students would learn the fundamentals of electronics in a not too academic way, and which should give them sound practical experience and a firm basis for further work.

Since I was already running a Keller plan course*, I appreciated the need for students to know clearly what it is they are supposed to be learning. I therefore provided a complete set of objectives for the unit lab, intended to encourage students during private study, and to explain to them the significance of the practical work.

The overall aim was to be able to design and build working versions of a variety of useful electronic circuits, starting with a set of basic one-transistor units. It seemed important to aim for the mastery of at least some of the work and to aim to develop confidence. In addition, I felt that coverage was less important than using the material as a vehicle for important concepts such as input and output impedance, and for skills such as use of an oscilloscope, soldering, and selection of component values.

The lab was designed to last for three weeks and to run continuously for one and a half days each week. To make close personal contact possible, classes were restricted to about eight students, running the lab several times in sequence. Informal group discussions could readily be conducted; using them for pooling practical data and knowledge. Thus a policy of getting the students to do different but related tasks could be adopted. Each week, a sequence of practical work followed by general discussion, instruction and related homework (including preparation for the following week) was used:

----

*see the project's companion volume, Individual Study in Undergraduate Science.

Week 1

practical work:	Initial instructions. Basic practical skills: soldering and component recognition. Building and measuring performance of a basic one-transistor circuit.
discussion:	Pooling of data. Model of a transistor.
homework:	Students study handout, objectives, and a chapter in the text book.

Week 2

practical work:	Construction and test of two of the seven basic one transistor circuits.
discussion:	Pooling of data and interpretations, this time about different circuits. Discussion of misunderstandings.
homework:	Learning the operation of all the basic circuits. Choice of project for week 3; start on design for it.

Week 3

practical work:	Mini-projects.
discussion:	Each student describes his project circuit, its intended mode of operation and how well it worked in practice.
homework:	Revision.

Assessment	Assessment is entirely by a short written test a week or two after the laboratory is completed. The questions attempt to put the student in laboratory situations.

One purpose of the mini-project carried out in the final week of the lab was to help students acquire confidence in their own abilities, and to some extent to prepare them for their third year project. Students were required to make use of the one-transistor circuits encountered before, to reinforce that work.

I could not myself be available continuously in the laboratory, and was forced to employ graduate demonstrators. The best role for me seemed to be to start the lab off, to give one or two demonstrations, and to run the discussions; the role of a graduate student was to assist students with their individual experimenting. These complementary roles meant that he used his talents appropriately, and I made the best use of my time. The development of definite roles, and the existence of objectives and other notes written for the students helped the demonstrator to see clearly what was required and what sort of teaching was needed, while leaving him considerable freedom to teach in his own way. The arrangement has worked well with a considerable number of demonstrators.

Students seem to have liked the integration of theoretical and practical work, both applying theoretical knowledge to practical situations, and constructing theoretical models from practical data. The lab seems to build up their confidence. Finally, they liked being taught by graduate demonstrators, not too much older than themselves. The success of the lab owes much to the enthusiasm and effort put into it by the postgraduate demonstrators; this in turn being due to the high degree of responsibility they were given and to the fact that their teaching role was well-defined (Davies and Penton 1976).

After the first two years the laboratory was transferred to the first year, with students now attending it for one day a week for four weeks. These changes made surprisingly little difference to its effectiveness or popularity, indicating the adaptability of the format.

Finally, recently I had to withdraw most of my own teaching commitment. The lab was put in charge of two graduate students who were now expected to run everything, including the general discussions. I gave them information as to the main points to be brought out in discussion and some general guidance. Even with this change the scheme seems to have worked quite well.

With these changes the lab has changed considerably from its original form. I now see the unit lab much more as a convenient organisational method for making proper use of graduate demonstrators and providing them with an excellent opportunity to do some good teaching. The scheme can bring out the best in all personalities involved.

6.6   COMMENTARY AND CONCLUSIONS

The point has previously been made (4.6) that it is just in subjects like electronics that it has frequently seemed natural to teachers to arrange for integrated teaching of theory and practical work, leading naturally to laboratories which can to some extent be compared with the unit laboratory described in section 6.4.

The comparison suggests, however, that there may be a number of topics, some yet to be invented or named, which could be taught in a structured sequence of varied activities. The structure of the electronics laboratory in 6.5. is based on a straightforward progression from simple to complex circuits and combinations of circuits. The structure of the electrical measurements laboratory in 6.4 is derived instead from ideas about how students might gradually come to terms with a whole class of complex instruments and their uses (it does not progress from simple instruments to complex ones, but from simple exercises with instruments to complex problems of using the instruments).

Another topic which is taught often by practical means, and which in the same way goes from simple to complex in activity, not in content, is computer programming. Again, it is not un-common to find short 'laboratory' based courses in this topic.

The teaching pattern at Liverpool in the first part of the first year, described in chapter 8, which involves the teaching of electrical instruments and ideas about errors, has some parallels with the patterns described in 6.4 and 6.5.

A marked difference between the accounts in 6.4 and 6.5 is the view taken of the ease with which such a laboratory can be handed over to somebody else. Much here depends, it would seem likely, on the subtlety of the thinking which has to be got across, and on the extent to which the work of the laboratory would immediately seem to a majority of colleagues to be obviously appropriate.

The four laboratories share a certain clarity of vision about what their originators want to achieve; a clarity no doubt to some extent induced simply by starting more or less from scratch. Attractive as such clarity may be, it cannot be taken for granted that it will produce better results than will some-thing which grew, evolved, and adapted itself. Indeed, in each case, the authors report adaptations and changes. At the same time, each laboratory described does at least take the risk of making clear what would constitute failure, with the result that in each attention is paid to evidence of success and failure and action is taken as a result. In a laboratory which has existed for many years and has passed through many hands issues of success or failure can sometimes become decidedly blurred.

REFERENCES

Black P J, Whitworth R W (1974) 'Unit laboratories in the department of physics at Birmingham University', in Studies in Laboratory Innovation, Group for Research and Innovation in Higher Education, The Nuffield Foundation.

Boud D J (1973) 'The laboratory aims questionnaire: a new method for course improvement?', Higher Education, vol. 2, pages 81-94.

Davies E R, Penton S J (1976) 'An electronics unit laboratory', Physics Education, vol. 11, no. 6, pages 404-409.

O'Connell S, Penton S J, Boud D J, (1976) 'A rationally designed laboratory mini-course', available from I.E.T., University of Surrey.

Squire P T (1974) 'Quasi-programmed laboratory demonstration: electrostatic field plotting', Physics Education, vol. 9, pages 45-47.

7. Projects

7.1   WHAT HAPPENS IN PROJECTS?

Project work is by now well established (Chambers 1970). The following portrait slightly exaggerates how it seems to have become established in a number of departments.

Five years ago, physics staff at X were perturbed by final year students' grumbles about their laboratory work, which consisted of a limited number of carefully planned and evaluated exercises, each having a comprehensive script detailing the procedure to be followed.   Experiments occasionally came and went, but the structure of the laboratory had been the same for about eight years.   The complaints were familiar and predictable: dullness, un-imaginativeness, irrelevance, and lack of scope for initiative. The timetable was said to be 'suffocating', and the whole pattern was said to lack awareness of the developing aspirations of final year undergraduates.

Attendance regularly sagged, making periodic exhortations necessary.   Cheating was occasionally suspected; staff discussions of it revolving around the poor motivation and lack of appreciativeness of the modern student.   Of course reactions varied: some who had demonstrated in the laboratory showed little surprise, even sympathy, but most either blamed the poor attitudes of students towards what was regarded as necessary rigour or as inevitable hard slog, while some took the chance to restate their long-standing doubts about the value of all practical work.

Gradually, encouraged by an unrelated decline in student numbers, opinion shifted towards favouring some response. Piecemeal change was rejected as unworkable, not least because several strongly urged changes were mutually inconsistent.   As something of a last resort, but not un-influenced by what was known to be going on in other departments, it was agreed to scrap the entire laboratory, while squeezing some of the more acceptable experiments into the second year.   In its place, final year students were to undertake individual experimental projects under the supervision of staff. The time allocation, and the contribution of assessment to the degree, were kept the same. One member of staff was made responsible for the organisation: for collecting topics, allocating them to students, and so on.

The change went smoothly, with only minor problems in defining ground rules. Staff seemed to know what was expected of them, and generally enjoyed thinking up topics, often related to their research. 'Something like a small piece of research' would fairly represent the generally agreed but rarely discussed definition of what a project was to be.

Four years later, the impression to be gained from talking to students and staff was that the change had been a good one. Complaints were few, working after hours was quite common, and most students showed enthusiasm for their work and pleasure in it 'being their own'. There were worries over lack of progress, especially when apparatus did not arrive or did not work, but supervisors usually managed to overcome such problems, aided by the fact that at least some students helped each other. Nearly every project turned out to be some sort of success, however modest.

Staff had come to see projects as an important part of the course, perhaps a focal one, justified both by their success and by their obvious validity as part of learning to be a physicist. Research benefited in small ways, some projects even being published. Knowledge of students gained when supervising their projects was felt to be genuine and significant, and was called on for writing references and giving advice. A few muted discordant notes included one from a supervisor who had found it hard to get on with several of the students he had supervised, and from another who felt that supervising projects ate too much into his valuable time.

The system had run in the same way since its inception, and had survived a new head of department with rather different ideas. There was no pressure, and there were no plans, for any further change.

The portrait is a composite of some of what was found by visiting physics departments, with a view to understanding better how project work functions in intended or unintended ways, and to seeing if there might be ways of modifying or extending it.

The investigation sought to interpret or make sense of what could actually be seen happening, and of what feelings and thoughts staff and students had about projects. Others have written of the aims, organisation, and assessment of project work (edited Goodlad 1975, Harding 1973) and this is not the place to rehearse those arguments again (the Goodlad volume gives an extensive bibliography). Rather, this chapter describes what seems to happen, and tries to reflect on it in the hope of raising issues which might bear consideration in various particular circumstances. Because each project is unique, and because they have only rather general aims realised (if they are realised) in a multitude of ways, general prescriptions are out of place. However, a comparison of different institutions can show up differences or similarities which deserve thought.

In so far as the above portrait rings true, it does appear that the functioning of project work in different institutions has something more or less common to all about it. The things which seem most common are:

projects are popular with both students and staff.

a project must reach a definite goal to count as complete.

initially, problems are proposed by staff.

students work singly, each with a supervisor, who usually proposed the problem.

the model in the minds of staff and students, which helps decide the nature of problems and the activities expected, is one of research.

the goal can be shifted, usually by the supervisor, in the light of how things turn out, so as to ensure that some goal is reached.

a final output in the form of a written report, sometimes supplemented by oral presentation, is required.

the report is assessed, the supervisor playing the major role in this process.

the assessment makes a significant contribution to the final degree.

the pattern of work is a change from that of previous years for students, in time scale (perhaps up to six months), and in its demands for such things as building apparatus, writing reports and so on.

supervisors judge projects mainly by the quality of the report and the achievement of the goal.

supervisors' day to day concerns are mainly with immediate practical progress.

project work is a stable pattern, not much subject to re-examination.

These and related features represent a frequently found grouping, and build up a picture of a common way in which that ill-defined notion, a project, is shaped by the expectations and constraints of the university. It is not the only form in which work with many of the virtues of projects can exist. Some, for example, have found ways of involving undergraduates more directly in actual research, sometimes with a very flexible commitment (Cohen and MacVicar 1976, Anderson et al 1974, Breslin 1974). Others have introduced the group study of a topic in depth, including project working, but also including activities such as giving papers, reading the literature, and so on (Black, Dyson, and O'Connor 1968 - see also 4.6). Some do not accept the view persuasively argued by Morrison that less may be more (Morrison 1964) and keep a pattern of a greater number of experiments so as to enlarge students' experience of equipment and techniques.

It therefore seems worth asking whether all the features frequently found in the normal pattern of project work are necessary or desirable, to ask how they may be connected with the educational virtues of projects which get most emphasis, and to ask which virtues may get less attention than they deserve, perhaps because the success of projects and their stable form distracts attention from them.

The investigation discussed here was intended to approach these questions. One person visited eight physics departments in British universities, interviewing or having discussions with fifty-five students in thirteen year groups, and also with a good number of supervisors. Most interviews took place during the projects, though some were held afterwards using diary material as well. They were informal and unstructured, students being encouraged to talk freely and confidentially about any aspect which they thought relevant. Three-quarters of the students were in their final year, this being the main focus of the study. Students, staff, and universities were assured of anonymity; their enthusiastic co-operation is gratefully acknowledged. The investigation was modest in scope, intending more to identify or clarify issues than to attempt to answer many questions.

7.2   THE VALUE OF PROJECTS: SUPERVISORS' VIEWS

Supervisors' views on the benefits of project work are naturally varied. They are also largely conventional, whether dismissive,

> 'We all know they are good for the students - anything that is practice at the real thing must be - so any attempt to identify individual benefits will be to some extent phony.'

or unexceptional,

> 'It teaches them to use their initiative and not give up when things are going wrong.'

or sometimes extreme,

> 'Students must be made to strive for a complete solution of the problem, not just elegant attempts that don't work properly. After all, life is about getting the right answer in spite of all the difficulties.'

Overall, views fell into three main areas. One area concerned the student's development as a person:

> 'They learn the value of tenacity - very little they know ever fits the problem in hand without some compromise, some adjustment, some approximation.'

> 'They learn to co-operate with others, and communicate too. They must learn to articulate their ideas if they are going to get the facilities they want.'

A second area concerns mainly knowledge and understanding:

> 'A problem to solve is an excellent stimulus for getting to know one relatively small area in depth. Then my students start to see the things they can do with their physics.'

A third area concerns more specific skills and techniques of scientific investigation, and the associated skills of writing, searching for information, and so on:

> 'They have not learned the proper approach to planning an investigation. Until they try it they don't know what sort of information they should put together before getting started.'

'How to write a full length report is very new to them.'

Individuals varied in the emphasis they gave to each area, but not in a way which suggested that their preferences were based on any strongly coherent analysis of what projects might be good for. Instead, they would often speak in analogies, saying for example:

> 'When you learn to drive it is important to go out for a drive somewhere. It is the same sort of obstacle course, and every one you get over makes you a little bit better. It doesn't matter where you drive to - it's just that you should go out somewhere.'

This analogy conveys that what matters is experience; that projects work by putting the student through a particular kind of process. A colleague disagreed:

> 'The analogy is invalid. It implies that the target is unimportant, and that is not correct. With a project there is the possibility of, if you like, only one journey, only one set of mistakes... Besides, the instructor is always with you and the penalties for making the wrong decisions are nowhere near so catastrophic.'

It seems likely that such relatively cloudy formulations work well enough for supervisors to conduct project work to their own satisfaction and to be sure of what they are doing, partly because a project has a clear focus in a definite problem, and partly because the activity is assimilated to something else which the supervisor understands very well, namely research:

> '... research scaled down, meaning that the sort of training that post-graduate research is considered good for can usefully happen in a reduced form, but to every-one, not just to those who carry on after their degree.'

Supervisors, then, saw the issues in general terms, rather than as related to the specific properties of particular projects.

Perhaps the strongest impression of the way supervisors saw projects was as a means which allowed and encouraged a great variety of events different from those which happen elsewhere in a course. These events could encourage any number of qualities, which would not be brought out by other practical work, because of differences in time scale and in the nature of the task.

Projects were also valued for various side effects, from

> 'A chance for the chap who is a chronically bad
> examinee to restore the balance a little.'

to, more generally, a boost for morale and leading to improved
attitudes towards the subject. One supervisor put his finger on
their potential value to a department as a diagnostic tool:

> 'A useful way of finding out the sort of area of teaching
> that needs more attention.'

When benefits are described in general terms, often by
analogy, it is hard to know if people whose views appear to be
in conflict actually act in conflicting ways. Thus one member
of staff in charge of projects as a whole in a department saw
them firmly as beneficial in allowing students to come into
intimate daily contact with the experience and judgement of a
research worker. Two others supervising projects under him
saw it differently. As one put it:

> 'It's about independence, and seeing jobs through
> properly from start to finish. I'm here to see that he
> doesn't do anything damaging - if he wants information,
> there's the library.'

It is possible - even likely - but by no means certain, that
students supervised by these people will undergo substantially
different experiences. A department in which such differences
exist, and they are not unusual, might do well to ask what
supervisors actually do, and whether or not that reveals a
consensus not apparent in differences of opinion at a general
level. It is even hard to know if those who take an extreme or
limited view of projects mean what they say. It may be doubted
if a supervisor who said that the major purpose was to get
students to write coherent and accurate reports, or if another
who said,

> 'a way of showing students how to negotiate the mine-
> field encountered by anyone who contrives to apply
> lecture room physics to the solution of real problems,
> and nothing else.'

acted in a way markedly different from anyone else. They may
have, or they may not.

Overall then, supervisors see, usually in very general, often
metaphoric terms, a great variety of useful outcomes of project
work on many levels. There is, though, little approaching a
general consensus which might lead to a set of agreed objectives
well enough defined to serve as a basis for making judgements

about whether projects succeed in achieving specific aims. At
the same time, the fact that the system can accommodate such
diverse views is one reason for its stability. Further, the
degree to which practice shows a greater definiteness than do
views about it, suggests that the actual situation of project
supervision leads supervisors to act in ways which only partly
reflect their predilictions. Nevertheless, it is still likely that
a student would find one supervisor very different from another;
differences which at once reflect the valuable autonomy and
freedom of the university teacher, but at the same time could
be a source of difficulties if left unrecognised.

7.3   BENEFITS AND PROBLEMS: STUDENTS' VIEWS

Students' comments about the benefits they obtain from projects
square only to some extent with the views of supervisors outlined previously.

Thus for example, several saw particular value in having
discovered a great deal about the practical functioning of the
department; about who was who, what was to be found where,
and how to find it. Others valued highly skills like soldering,
which could be thought to have been things acquired rather
earlier, and had in the cases where they were mentioned often
been included in a previous workshop course. Even soldering,
it appears, can take on new meaning in the context of doing a
necessary job for oneself. The connection with the real needs
of the moment comes out well in comments like,

> 'The problem of finding a grease that had all the
> properties I wanted and which was also insoluble in
> toluene was enough in itself to make my project useful
> to me.'

In the main, students tended to agree with the 'learning to
drive' metaphor of the previous section. That is, they saw
benefits as deriving from the particular situations and difficulties
they had passed through, and as being greater the more of them
there had been. The contrast in this respect with traditional
experiments was often mentioned. Process, not product, tended
to fill the stage.

As with tutors, students took varied views, so that one
asserted that doing his project had taught him only to do a
project, and another loftily spoke in terms of working at the
research front of a field.

138

Students were asked what gave them pleasure. Almost all mentioned the moment when they first began to get some definite results:

> 'It felt good when I started to get out some equipotentials - until that moment there was no certainty that I'd have anything to write about.'

> 'I felt great when I found the maths had gone right and the curves fitted my data.'

Clearly this moment was a watershed looked back on with pleasure perhaps out of proportion to its overall importance in the whole project. By comparison, finishing the report or getting a good assessment featured much less strongly.

Replies to a question about sources of frustration were varied and coloured, often with concrete detail. Equipment availability featured strongly:

> 'I've been waiting for weeks for a simple vibrator to arrive. I hope they will take this into account, but I don't see how they can, not properly.'

Here the associated anxiety links with assessment. The effect can also be on interest and drive:

> 'We now find there are some ready-made filters on the third floor we could have borrowed instead of making our own. It diverted attention from the real problem, and we lost a bit of interest.'

Equipment is often the focus of trouble in getting the system to deliver what one wants, or to allow one to do what one wants:

> 'I wasted time looking for a suitable oscillator... I had to make one eventually. I imagine there is one some-where, but we weren't encouraged to go and look.'

> 'The steward from the first year laboratory took the lens back each week.'

> 'It's hard to describe to a technician what you could use as an example.'

> 'It takes so long getting things made - stirrers and so forth - that you give up really. There's no scrap lying around to do a bit yourself.'

'Costs were always stressed when we thought we had a good idea.'

Those supervisors who see value in overcoming such practical hurdles can find some support for their views here, though they may need to recognise that the height of such a hurdle may be an accidental and arbitrary thing, out of proportion to the value of overcoming it.

Not surprisingly, students blame inadequate equipment. Some may be justified, but of course it may be a question of the poor workman blaming his tools. Lack of knowledge or uncertainty about how to get it were also mentioned by not a few:

'We did a literature survey and found nothing relevant - perhaps we should have tried again.'

'I spent ages finding out about glass, but it was no use because I couldn't find what sort of glass they had given me.'

It becomes clear from these and similar remarks that the differences between 'good' and 'weak' students thrown up by projects are varied in the extreme, including the ability to wheedle materials out of a technician, to think of ringing a manufacturer about the drying time of his paint, to think of ways of pushing modest apparatus to its limits, to know when to try the library and when to ask, to tolerate frustration and boredom, and so on through a long list of items which have in common only that they can sometimes happen when one tries to do a practical job of work. As one student saw it:

'It's just an obstacle race - you hope you are still jumping when June 8th comes.'

The question is whether the obstacles are essential ingredients deriving from the real nature of the task - a view which can be used to justify anything which happens - or are sometimes unnecessary barriers to the completion of a satisfactory and satisfying piece of work.

An analysis of replies by nineteen biology students doing major projects in a degree course to a question asking them to list words describing the nature of their experience of projects casts some light on these reflections (Thomas 1976).

Two thirds of the 125 words produced were emotional reactions, equally divided between positive and negative. All but one student included words of both kinds, usually about equally balanced. The largest single group of positive reactions was the set 'interesting, stimulating, exciting', with the four groups 'interest', 'involvement', 'enjoyment', and 'fulfilment' accounting for eighty percent of positive reactions, the rest being classifiable mainly as 'enlightenment'. The largest single group of negative reactions was 'frustration, exasperation, annoyance', with the rest accounted for in four equal groups 'worry', 'tension', 'tiring', and 'boredom'. The most striking feature was the way these opposed reactions appeared together. Typical lists were:

exasperating	frustrating	exciting
tiring	interesting	exhausting
inventive	experience	frustrating
enjoyable	stretching	revealing
interesting	satisfying	despairing
fulfilment	worrying	stimulating
	annoying	all-involving

In a word, apathy is not what these answers reveal.

Returning to the investigation of physics projects, students' reactions to supervision again show considerable variation. More supervisors themselves decide the amount of contact than leave it to students, and more opt for frequent than for infrequent contact. A guess at the working time between contacts might be two to three hours, but can be less in special cases:

> 'My equipment is in my supervisor's research lab - it's really a spin-off from his main work, and he is there seven-eighths of the time, so he checks on my progress all the time.'

Relations with the supervisor are not unimportant. Students want both to be helped and to be left alone, to be told and not to be told. So one will say,

> 'They just nod and say that's OK, and when it doesn't work they suddenly find out what what they said was OK wasn't really.'

while another will demand independence:

> 'We were perfectly capable of working it out for our-selves. You really ought to have every chance to show

what you can do.'

Not all see, or are encouraged to see, a project as their own, but expect, or are led to expect, the main decisions to be made by the supervisor:

'We worked on an electrical way, and got it partly working. But he came round and said that wasn't allowed - it had to be a mechanical method. He obviously had a method in mind and we hadn't read his mind correctly.'

The supervisor has a difficult job, and naturally some do it better than others:

'Choice of supervisor is more important than choice of title. You can at least bend the subject a little.'

One student reveals both sides of the coin:

'It is so frustrating having a supervisor like mine. He knows so much that I think he could do my project in three hours flat. But he is kind enough to avoid giving me that impression too much.'

while another illustrates only one side:

'When we said what we thought was worth doing, he just added more problems all the time until it became impossibly complicated.'

Choosing a topic is another illuminating aspect. Topics seem to be chosen more often for unusualness, usefulness, and special interest rather than for the help they might offer with understanding subject matter. So topics like measuring noise in cars, measuring heat output from the human body, making a wind-driven generator, or designing a fuel flow meter were attractive to many. One, however, saw it as a special opportunity:

'I wanted something really formidable to test my suitability for research.'

This student clearly values the process, while the next at best puts up with the system:

'Mine was chosen by elimination of degrees of boredom.'

However, there were few signs that choice was critical,

another indicator that it may be the process which matters more than the product. Even those students who forgot, ignored, or missed the publication of a list of topics did not seem to suffer much, and nor did those who got only their second or third choice. Such an effect may be related to the fact that students cannot know much about what they are in for, unless work in previous years has had a project ingredient.

Many of students' views about organisation were clear. A chance to work outside normal hours was valued. They found it helpful to be required to give some kind of interim progress report, written or in the form of a mid-period colloquium. Again, a requirement to produce a feasibility study met with support. It seems that organisational arrangements which help to structure effort and planning help the student, and probably the supervisor.

Location can make a difference. Working together in a special project laboratory, a number of students made a point of the value of collaboration:

> 'One was good at electronics, another had a flair for computing, a third had worked in an industrial laboratory and knew about PTFE and that sort of thing.'

Where work was based in research laboratories, students could not easily help each other, but some valued the constant access to professional advice.

It might be supposed that report writing and assessment would bulk large in students' minds, but it does not appear to be so. The report seems to be regarded as a necessary job to be done, though it is a help to see examples from previous years. If anything, attention centred on appearance and length, rather than on more imponderable things like the actual quality of the work. Again, in spite of considerable variations in methods of assessment, fairness was not felt to be much of an issue. Some favoured an extra oral examination, as helping those who found writing difficult, with an oral given by an outsider who did not know anything about the project finding most favour. As reported elsewhere in the book for other practical work, details of the assessment method and mark breakdown were normally not common knowledge. Little conflict between the roles of supervisor and assessor was perceived. Whether one had an 'easy' or 'hard' project seemed mainly to be thought of as the luck of the draw, and difficulties and frustrations were not so much seen as a comparison of one's progress with that of others, as judged in relation only to progress towards one's own goal.

Many students felt that some preparation for projects would be helpful. Suggestions included their introduction into earlier years, and having two shorter final year projects with only the second being assessed.

Despite students' involvement in and attention to their projects, other influences do have some effect. Clashes between the demands of final examinations and projects were often cited. Pressure arose too from the (also valuable) close attention and interest of staff, so that some claimed that they had economised on course work to meet this demand; a demand the stronger because concealment of inaction or ignorance is harder:

> '... unlike examinations, where with a little skill you can avoid showing them what you don't know.'

7.4   QUESTIONS AND COMMENTS

Some of the impressions accumulated gradually in the course of visits to departments are rather more general and less tangible than most of the foregoing. They are interpretative and fallible, but may nevertheless provide food for thought, even to those who believe them to be mistaken.

First, there is an impression of what might be called the 'popularity syndrome'. On the whole, project systems run smoothly and are popular. It seems paradoxical to suggest that this may conceal ،a danger, but one impression left by the visits is that the very popularity of projects has the danger that little questioning or critical thought is given to whether their potential values are in each case being realised. Staff clearly think about projects on two levels which are some distance apart: the general, often metaphoric level documented in section 7.2, and the level of daily practical difficulties. The gap between could contain more thought about the value of the particular processes and experiences generated by each project, perhaps for example considering the relative value of accumulating large amounts of accurate data as against planning or construction. Few supervisors seem to choose a topic on the grounds of the precise kind and range of experience it might be hoped to call forth, or to recommend a change of direction in a project on similar grounds.

In so far as this impression is correct, there is the further impression that there are reasons for it, having to do with the nature of the normal pattern of project work described in section 7.1.

One reason could be called the 'success syndrome'. That is, supervisors appear to feel some pressure to make sure that 'their' projects succeed in some sense or other, perhaps partly because of a wish to compare well with other colleagues, and partly because of a natural desire to make sure that the student reaches a satisfying conclusion. In practice, what seems to happen is that the goal of the project is continually subject to revision, so as to ensure that some target, however small, is reached. So, when time runs short, for example, a student may be told that he had better measure the heat capacity at one temperature, rather than over a range as planned. If the planned material to be investigated is destroyed, lost, or does not arrive, some sufficiently similar material is identified. In this way, every project is a success, the difference between good and not so good being the size of the success.

This means that some aims expressed by some supervisors are achieved more in the letter than in the spirit. 'Seeing the job through to the end' may have a slightly notional meaning in some cases, for example. The consequences of adopting a different policy would be considerable, and not lightly to be undertaken. The nature of assessment would have to change, and might have to concentrate more on the things the student had done, including evaluating the merits of his mistakes, as opposed to weighing his final product, which might not exist.

A related reason might be called the 'product fixation' syndrome. This is based on the impression that supervisors tend to see the project in terms of the outcome, often as embodied in the report, and not in terms of the processess leading to it which students emphasise strongly (see 7.3). Thus things which contribute to a good report may be overvalued against things, like a bright but misleading idea, which do not.

Rather deeper, perhaps, is the impression that there is a 'reality' syndrome. Projects are universally justified in terms of their approximation to reality, but it may be that the nature of the approximation matters quite as much as what is approximated to. Unexamined, 'reality' cannot be a good justification, because it justifies anything. Thus, if a project involves twenty hours of patient data collection, it is said that that is what life is like. If it involves much machining of raw metal, that too can be justified on the same grounds. But, despite the validity of such a view, there must be limits beyond which it cannot be pressed, if only because a project is not real life, but is rather an ingredient of a course which has some useful properties deriving from the actual demands of a situation. Some aspects of real life, notably dismal failure, are rightly well protected against.

Another problem of using reality to justify anything is that doing so can distract thought from the relative merits of the various things the student is called on to do, as mentioned previously. There may be, then, value in further thought related to individual circumstances, about how far and in what sense a project is a model experience. One might consider just what parts of the work likely to be required by a particular project have greatest and least value, and modify what is given to or required of the student accordingly. None of this is to say that this is never done. In a general sense it is always done, but there may be scope for doing it more.

A rather firm impression is that in most physics departments, research is the model by which projects are judged. The model is especially attractive if the department sees its main job as being training for research (or argues that that is what it knows best). Miniature research can no doubt be justified as part of the education of a teacher of physics in schools, or of a manager, but the argument has then to be stretched a little further. Some departments are more ecletic than others in the kinds of project they encourage. While one will place the highest value only on topics that, well done, could lead to publication in a scientific journal, others encourage such things as doing a job for a local firm, making films of experiments for use in the department, and so on. Widening the scope is not simple. When one student wanted to make a psychedelic lighting system, the department only resolved the debate over whether to allow it by insisting that it must be capable of producing patterns appropriate to Mozart as well as to rock. The issue of scope may nevertheless bear examination.

Supervisors have to decide what and how much help to give, and usually do it on an individual basis. While the amount of help given varies, it is usually, it seems, a middling amount intended more or less to ensure success without destroying independence.

The student is rarely abandoned to his fate, or led by the hand, and most would strongly reject doing either. Both alternatives can, however, be put in a more favourable light. If one did not insist so much on the value of the research model, or on the need for a clear and successful outcome (both derived by analogy from something outside the project scene and not necessarily inherent to it), it might be possible to envisage setting rather vaguer tasks which lent themselves to development at different levels of complexity or of depth, and gaining more of the goals of independence and initiative than some projects currently do. Equally, there could be value in seeing the project as much more a matter of working day by day at the shoulders of people doing research, with a constant interaction which would in a conventional project system look like 'help' in large amounts, but there might just be natural.

To say this is not to urge change, nor to press for change in one direction or another. Rather, it is to say that the way project work is currently made to fit smoothly into the system makes actions, which from another point of view would look the obvious thing to do, seem unreasonable or impossible. It is to suggest that project supervisors might profitably ask whether there are ways of widening the range of what projects achieve, or of focussing on fewer aims which they feel to be more important than others, instead of accepting the product, accidental or intended, of the way things happen to run at the moment.

In a more particular spirit, there may be special outcomes of project work which a shift in emphasis could encourage.

One supervisor reported in section 7.2 said how project work was a sensitive mirror of the success of the rest of the course. Potentially, he could be right, but it is worth noting there where the supervisor intervenes so as to ensure success, this mirror is at best misted, especially when all intervention is private.

Some attach great value to projects as encouraging communication and co-operation, but in general most seem to accept what happens to emerge, rather than deliberately stimulating it. If one does believe strongly in its value, and students do not at present work together in one project lab, do not give oral presentations, or if they invariably work on individual unrelated topics, there may be scope for change. Even working in groups, usually rejected for reasons of assessment, but nevertheless practiced in some institutions, might be considered (in one institution, groups decide themselves how to share out the total mark given).

Others have argued for the value of projects in revealing unique qualities and in compensating for weaknesses which are over-emphasised by written examinations. For them, there might be merits in reducing the closeness of supervision, especially where it pushes the project in a direction the supervisor feels right.

Yet again, others stress the project as an opportunity to know a field in depth. Here on could argue for more supervision rather than less, since 'depth' involves a whole complex of kinds of knowledge probably best taught through frequent personal interaction at several levels.

There may also be some ways of increasing the reasonableness of the job presented to the student. Departments usually provide reasonable facilities, but some make it easier for students to get technical jobs done, or borrow equipment, than others. Some make a virtue of 'throwing them in at the deep end', while others give at least some thought to preparation along the way. Some give students the freedom to buy small items, obtain outside information, get access to advice from other specialists, while others do less in these directions. Some limit firmly the type of topic, while others give more freedom of choice. Some take in reports and issue marks, others use oral examining sometimes with the student in front of his apparatus.

Of course, it may be that a supervisor feels on reflection that projects run as he presently does achieve about the right balance of aims. What it may be well not to take entirely for granted is that they are certain to.

The practitioner, especially the man in charge of a project system, may well smile ruefully at all these reflections. He knows only too well how the real business consists of extracting topics from tardy colleagues, of dissuading students from their flightier fancies, of coping with the supervisor who suddenly has to leave to join his research group elsewhere, and so on. As much as anything else it is the division of responsibilities which determines the character of project work, a division which strengthens the system by allowing everyone to do more or less as they want, but which makes it hard to discuss or implement any change in policy. A first step in the direction of thinking about policy might be to collect some sample reports and discuss them, asking if these students got what they might have from the experience, or were encouraged to give what they might have. Such an exercise might confront the metaphor of the project as a 'real' task with the realities of what happens in them, and out of it might grow a clearer picture of the relation between the two.

REFERENCES

Anderson W F, Madjid A H, Pedulla J, Martinez J M (1974) 'An avenue of undergraduate involvement - research', American Journal of Physics, vol.42, page 944.

Black P J, Dyson N A, O'Connor D A (1968) 'Group studies' Physics Education, vol.3, page 289.

Breslin L (1974) 'A student's view of undergraduate involvement in research', American Journal of Physics, vol.42, page 948.

Chambers R G (1970) 'Laboratory teaching in the United Kingdom', in New Trends in Physics Teaching, vol.2, UNESCO Paris 1970.

Cohen S A, MacVicar M L A (1976) 'Establishing an undergraduate research program in physics: how it was done', American Journal of Physics, vol.44, page 199.

Goodlad S (ed.) (1975) Project Methods in Higher Education, Society for Research in Higher Education.

Harding A G (1973a, b) 'The objectives and structure of undergraduate projects' and 'The project: its place as a learning situation', British Journal of Educational Technology, vol.4, page 94 and page 216.

Morrison P (1964) 'Less may be more', American Journal of Physics, vol.32, page 441.

Thomas M (1976) questionnaire data privately communicated.

8. Progress reports 2

8.1   INTRODUCTION

The three accounts of laboratories which occupy this chapter
have in common that they raise, in different ways, questions
about how innovations in the laboratory can succeed in fitting in
with external constraints.

The first two accounts are of very different attempts to
rethink in a radical way the content of a first year laboratory,
though both share a view of the laboratory as a course in its
own right. They differ in what they regard as appropriate content, and in how teaching is structured; differences which owe as
much to circumstances as to philosophy. They also differ in
scale.

The third account concerns, not the detail of any one practical course, but a pattern of organisation, termed unit laboratories, intended to make it possible for various members of
staff to work out and implement their own ideas. It shows how
the electrical measurements laboratory described previously (6.4)
is by no means the only possible form which can fit within the
scheme.

8.2   DESIGNING A LABORATORY COURSE
AT ROYAL HOLLOWAY COLLEGE

This account is based on material by Susan Kay and Sid O'Connell, together with information collected from students by Garry
Dearden and Martin Harrap in the first two years of operation.

When Susan Kay recently took over responsibility for the
first year laboratory at Royal Holloway College, and in collaboration with Sid O'Connell and Simon Penton of the I.E.T.,
University of Surrey, considered reorganising it, they all felt
that its traditional circus format had several disadvantages.

There were plenty of experiments, categorised according to

subject matter, but without any obvious plan. The number of experiments in each category varied a good deal; for example it turned out that electricity was well represented, with some four or five experiments on the potentiometer alone. Experiments were assigned to the first year rather than to the second on the basis of the sophistication of their subject matter, with no planned development of skills over the years, except for the introduction of third year projects.

A second feature was that, as usual, experiments were allocated according to their availability, so that any one student's laboratory course was ad hoc, and might not be balanced. The experiments were all the 'same', in the sense that each took six to nine hours , all were formally written up with emphasis on data analysis and errors, and that nothing distinguished one from another except its subject matter.

Thirdly, with no one person having been in overall charge, experiments had remained largely settled, with of course new additions from time to time on individual initiative. Much of demonstrators' time was devoted to marking reports, both in and out of laboratory hours; a teaching activity of limited scope, value, and interest.

Encouraged in part by discussions with others in the project, and perhaps by a move into new teaching laboratories, they felt, after some initial trial of new ideas such as short group projects added to the laboratory, that they wanted to think through the whole structure of the laboratory more radically.

That they did just this is illustrated by remarks of students looking back at the rearranged first year in retrospect:

> 'Working in a group seemed more realistic. We sat down and discussed what we were going to do, and having to agree - although I suppose I didn't like it when everybody said I was talking a load of rubbish - made you more confident with each other, that you could say things and criticise each other.'

> 'We built something that worked as it should according to predictions - we achieved something.'

> 'I liked the way we were introduced to everything first - all the different instruments and techniques.'

Very clearly, these students are not talking about a year spent doing a series of individual experiments, but about a year spent doing a number of different things, including group disc-

ussion, project work, and training in techniques. Another intended feature of the change shows up in:

'It was good in teaching you how to use a lab, and what
to do in a lab, but I don't think it was particularly back-
ing up the courses we were doing.'

What is the programme? It is complex, and the activities
vary week by week. Some occupy an hour or two, while others
may last over two or three sessions. The programme has
changed in detail each year, often in the light of previous
reactions, but the following outline is typical.

The introductory sessions contain talks and some demonstrations which provide material for discussion.

Several succeeding sessions might be occupied by a mixture
of seminars and work devoted to learning techniques and skills.
The seminars concern such topics as systematic errors, or the
discussion of an experiment with a totally unexpected result. The
techniques, besides instruments such as the CRO and multimeters, include also such things as the use of logarithmic
graphs , the use of calculators, paper exercises in data analysis, the use of journals, and short exercises in scientific writing. The second term includes a short unit of electronics.

In other sessions during the first term or so there are some
problem-solving experiments, such as determining the sinking
time of a model 'boat' (Richards 1974), in which students decide
which variables to regard as relevant, and in what controlled
way to collect data to reveal their relationships.

Such relatively closed activities are offset by some short
open-ended investigations intended to encourage initiative, arouse
interest, and allow the maximum personal interaction.

As the year goes on, other activities build up towards an
individual project at the end. In this, students can choose
whether to exercise a good deal of initiative, or whether to work
on a more limited and well-defined topic ('measure the refractive index of air over three orders of magnitude of pressure
variation').

The activities include a good deal of team work. In one, a
situation is set up in which a group of five or so play the role
of a research team presented with a problem such as the design of a device for measuring the mass and surface area of a
hair. The groups report progress to a 'section leader', and to
'management', these roles being played by post-graduates and

staff. At the end, the groups report at seminars.

Such work develops in stages. Thus a group project in which the group actually has to build a device might be preceded some weeks before by a similar notional exercise.

What was the rationale for this structure? It is one in which different activities of many kinds are planned to achieve, or to work towards, goals of many kinds, both short and long term. As Susan Kay and Sid O'Connell write:

> The laboratory, free from the limitations of attending to one bit of subject matter at a time, seemed to us the place to attempt to teach those things which transcend subject matter. We came to feel that it was right to go a long way in this direction, not just planning to teach experimental skills, but also asking how we could plan activities which would develop such things as critical thinking, the ability to communicate, or the ability to work effectively in a team, not least because the lack of such qualities in recent graduates is a matter of some general concern. That is, we wished to make the laboratory more educational, in the sense of widening the range of things it tries to do.

In addition, we proposed to attack these things more directly and explicitly than is usual, our view here being well summed up by Ruth Beard:

> 'Teachers in tertiary education have a diversity of aims. They expect students to acquire a body of knowledge, to develop the ability to learn independently, to solve problems, to be intelligently critical of what they read, to express themselves articulately, and so on. Each aim needs to be matched by appropriate activities, teaching methods, and evaluation.'

An example of the way these ideas were followed up at Holloway is offered by work to do with written communication. Rather than tacking the writing of an account on to the end of an experiment, students are given a series of writing tasks, perhaps starting with describing events seen on a film loop, and later writing at greater length about an experiment of their own choice. The writing is discussed as such, being criticised and redrafted until a satisfactory version emerges.

The laboratory raises a number of questions, not all of them easy to answer.

What do students think of the various activities? The answer obtained by asking them is often tangential to the real issue, which is what they would have thought of some activity of the same kind, directed to the same ends, but perhaps better exec- uted, or done at a better time. That is, one wants to know how to do better, because nobody could hope that all, or even most, of such a set of varied activities would be got right first time.

So, for example, when initially some very brief open-ended investigations were included in the introductory sessions, students could be found saying things like:

'The thing that struck me in the first couple of weeks was the lack of anything really happening. I would have liked to have got into something, right away, the first day. Here we have all this equipment - it seems such a waste not to use it.'

'I didn't like the open-ended investigations. They seemed to drag a bit. I could see the point of them, but I seem- ed to run out of ideas.'

Information of this kind pointed to the likelihood that the inv- estigations were coming too soon, at a time when students did not know what apparatus they could get hold of, and that they were being presented with too many given ideas about what to do. A change to a later date, after work on techniques had as a by- product shown students the apparatus, and some changes of pres- entation, seemed to help a good deal. In such a case, then, it is arguable that the course can be improved by changing detail, not structure, despite considerable adverse reaction. Understand- ing the causes of reactions (good and bad) is essential, and of course very difficult.

By comparison, after a term many thought well of most of the exercises devoted to techniques:

'The techniques were the most vital, and the most enjoyable part of the practical course.'

with strongest approval perhaps going to things which they saw as having obvious, immediate, and definite point:

'You see things like AVO's everywhere, so to be able to use those and know their limitations - particularly to know their limitations - I can see being very useful.'

Students then in their second year looking back, mentioned techniques less often and intervening activities more. While not generally hostile, some could now be found to say things like:

'Nothing I've done this year has been helped by last year.
What I've done this year I've had to learn new - like the
use of equipment. I know in the first term we had to go
through just about every piece of equipment in the laboratory...but we didn't really get enough time. By the
time we needed to use them this year we'd forgotten
half we learnt.'

The problem is whether it is the special point about time
that matters, or whether the dismissive tone, common to much
said by second year students (though not all) has a deeper cause.
There probably is a deeper cause. Second year students look on
the first year laboratory course as isolated and different from
other work in the first year, and from work including laboratories in the second. They are obviously right to detect a difference, and it is not surprising if they feel the smaller and different component to be the aberration, rather than the other way
around. It may not be only for this reason that some felt the
first year practical work to be uncertain in its direction and
point, but it may well contribute.

The lesson, perhaps, is that it must be difficult to convince
students of the value of work whose assumptions run counter in
many respects to what they find at most other times. Changing
established ways of thinking is hard, and a new course cannot
achieve it quickly.

Such an inference gains some support from the nature of the
very strong approval given to the electronics unit in the second
term. Comments were of the kind, 'You really did learn something'; comments which may owe more to the highly structured
and easily perceived nature of what is learned then, than with
any great advantage in the effectiveness of the unit.

It gains further support from the way reactions to work with
unusual features, like group work and discussions, were strong
but divided. Those who did not like them were dismissive:

'The discussions were new. But you couldn't get anything
from them. It was so pointless. Working out how much
rubber is worn off a tyre in each revolution; what could
you get from that?'

On the other hand, those who reacted well could give good
reasons for doing so:

'It was useful in that I learnt to discuss my ideas with
other people and take their criticism...and having to
convince other people.'

'We'd discussed it so much that we could discuss it with others, and explain what we were doing, and why we were doing it, and ask them if we got stuck...It gives everybody a chance to do what they're good at. It makes you more individual.'

These students saw point in the exercise, and the former ones did not. The strongly favourable or unfavourable reactions were frequently tied to seeing the point in this kind of way. Those who did not see the point tended to take a more traditional view of the purpose of laboratory work:

'It didn't seem as solid practical as what we do this year. This year we can see we're doing things. Each week you do a new subject - you can say you've done that bit.'

No student took conservatism to the point of rejecting the project work planned as the culmination of the year, with the individual projects getting widely shared acceptance, even by those who regarded the second year as a return to 'normal'.

Overall then, a picture emerges of parts of the course (projects, electronics, problem-solving experiments, techniques) accepted by many, and of other parts strongly accepted or rejected. That the course was different, in good and bad ways, was often mentioned:

'A very good thing was that you didn't do formal write-ups...You wrote things up as you went along, like a diary, which was totally alien to anything I'd come across. So you didn't have "Procedure", "Apparatus"; rubbish like that.'

'That's the funny thing - last year they liked it done that way, but this year they want good write-ups. It's difficult to change.'

Conflict of this sort, as well as differences of opinion about how well parts of the course were conducted, may be responsible for the wide variations in general reactions:

'We were given quite a lot of freedom. You were more relaxed. The set experiments we have now would be a real barrier then.'

'It was something of a change, something different, something interesting.'

'There seemed to be no aim to it. We didn't seem to be

getting anywhere.'

'There wasn't a great deal of learning in the whole course.'

In any appraisal of the outcome, it is essential to remember the relevant circumstances. By contrast with the next account, which concerns an innovation conducted by a substantial group of members of staff in a large department, in a small department such as Royal Holloway it is unreasonable to expect more than one member of staff to play any very active role in just one laboratory. To attempt changes on the scale described here was courageous. Failure in some respects must surely be inevitable, and those who would deny anyone the right to be wrong make change in any direction difficult if not impossible.

A lesson here is the importance of collecting information about how things are going, and acting on it. As mentioned already, the first investigations were later delayed; they were also placed after problem-solving experiments which appeared to lead into them well. Discussions had tended to ramble, so they were made briefer. Group projects offered a number of difficulties, and so have been placed after individual projects, and limited to only a fraction of students. Continuous change and adjustment of this kind is an essential feature of the laboratory.

What is no less important is the nature of the change that was attempted. It was towards a much more active teaching role than is common, through a variety of different activities in the laboratory. In this, of course, the small department has logistic advantages over a large one (though see also the next two accounts). Those who are attracted towards similar ideas might learn lessons about what can be attempted, and about the ease or difficulty of bringing off various activities. Those who are not may find it useful to ask where they believe that they can or do encourage the development of similar qualities.

8.3   FIRST YEAR AT LIVERPOOL

The rationale and structure of the Liverpool laboratory is described in a two-part paper (Court, Donald, and Fry 1976) which argues out the basis of the laboratory and explains in detail how the ideas behind it are implemented. In outline, the class of eighty first year students, each attending the laboratory for two three-hour sessions a week, go first through an introductory term's teaching in groups of about ten, after which they undertake about six out of ten experiments derived from the original circus which previously occupied the whole year.

More is hoped for from them in these experiments than used in practice to be expected, especially by way of critical analysis of results, and of some modest measure of originality.

The introductory term has for this reason a course in data analysis as a major component; a topic which comes as near to being emotive as any in university laboratory work. This is one reason for discussing the Liverpool laboratory.

Another reason is that the paper mentioned above, clear as it is in its presentation of arguments and facts, is no guide to the wholeheartedness with which the teaching is approached in practice. It is our belief that this wholeheartedness has been essential, and that an attempt to reproduce the course without it, and more generally, almost any attempt at innovation without it, would fail. The Liverpool laboratory is a good one to exemplify the general point, because in it enthusiasm, hard work, and willing co-operation succeed in bringing off with a success it is hard to deny, an enterprise which on the face of it looks implausible in several important respects. Imitation of it, or of another innovation, without the hidden but crucial ingredients, would be perilous.

The peculiar place of errors and data analysis in thinking about laboratory work comes mostly from the fact that they cut across all measurement. The need, or otherwise, for a student to understand equipotentials or the Fabry-Perot interferometer can be decided in relative isolation, but what he learns about estimating random error might influence all his future experimenting. And although many would see this influence as something which fundamentally separates the men from the boys, there are others who fear the danger of a little learning in this field inhibiting inventiveness when it is inventiveness which ought to be encouraged. At the first year laboratory which is the subject of the first report in this chapter, it was decided almost to ignore the question of students handling errors, so as to increase their confidence in their own planning. This fundamental opposition of views is not the only problem in teaching about data analysis: beyond a limited basic area there is less agreement than one might expect on what is the correct understanding which could be sought.

The Liverpool staff, agreeing data analysis to be an essential first step, visited other universities to get ideas, and returned with the most satisfactory idea of all; that they could do better than anything they had seen. They tried a new scheme, modified it, used it several years running, and we saw its first term in operation after reading an early version of their paper.

158

All work in the first term is in groups of about ten students with one teacher to each group. Two weeks are spent on electrical measurements, and are not distinctly different from what might be encountered elsewhere. Two weeks are spent on mechanical measurements, the measurements being of L and T for a set of ten pendulums, leading to values of L/T^2 with a small spread. Four weeks are spent on data analysis, in which two experiments are the heart of the course; the measurement of a model representing an annulus from which a value for the volume of the supposed annulus can be calculated, and an experiment in which the masses of ball bearings apparently filling a measuring cylinder to various levels give an approximation to a linear relationship upon which the method of least squares fit can be practised.

The paper sets out in detail how the various objectives of the course are to be attained with this work. Apart from mentioning that simplicity of apparatus is an advantage, the paper does not openly make the case that the major objectives are the more thoroughly achieved through the elimination of lesser objectives, like enlarging experience of physical phenomena, although this motive may have been present. For example, while the counting of nuclear disintegrations could provide data as good as partly filling jars with ball bearings, many students might thereby become engrossed in nuclear irrelevances and have no attention left for the statistics. But it is hard to imagine that this danger could have been so thoroughly guarded against fortuitously as it has been in the Liverpool course: the jars of ball bearings are quite without meretricious appeal; the volume of the annulus, as distant from the main-stream of science as anything one can easily think of, is further deprived of reality by assuming that it must be circular; and, if there is a part of experimental physics with which freshmen have generally become fed up even before they arrive at the university, it is the simple pendulum. It cannot be argued that adequate experimental situations for exploring the techniques of data analysis can only be found amongst such petrified physics; if that were so, the whole case for data analysis would collapse.

Reading the paper before visiting Liverpool, we wondered to what extent students might feel that the staff had assembled some of the ingredients of rather a dull course. But our fears were not realised when, at our second visit, we asked for their opinions. One student, asked what he thought was best about it said:

'Hard to say - the best thing - perhaps working in a group. You can interact with other people. You're not working so much on your own... it's not what you might

159

define as work, you enjoy yourself more.'

Another, asked what was the worst thing, said:

'... it was the fact that I felt very lost at the beginning. Now I'm thoroughly enjoying it.'

Yet another student, who, when asked what it was like and what seemed important in it, replied:

'I like the way he explains things. I think he explains them well, and I can't really say there's anything we've done which I haven't understood.'

Not all students spoke of enjoyment. One, who was asked if he had done anything of this sort before, said:

'Not in that detail... I think it's a boring subject in itself, but the way it's been done isn't, you know, not too bad really.'

and, asked if he would have liked to use more sophisticated equipment, said:

'Yes, but then... if you've got to do this course on errors, then simple apparatus is the best thing to do it on, I suppose.'

One articulate student is worth quoting more at length because the view expressed would have been fairly widely accepted:

'... I suppose it did get a bit slowish but it wasn't too bad because the whole concept was a new one - actually working out an error in such a way. It was quite interesting as long as you were following it through. I mean there were some stages where you're just writing down a lot of figures and getting a result and then understanding afterwards. But for the most part you're understanding it all the way through and you're aiming for a point and you realise you've got to get there in the end, so...'

'What about when you're not in that state?'

'Well then it can get monotonous... I suppose it's the same with anyone. If you can understand something and you can see where something's leading, it's going to be interesting even though it is monotonous. If you don't understand it and it's monotonous, then the fact that it's monotonous becomes more apparent.'

This more apparent sort of monotony had been systematically rooted out. Research students had been thought incapable of the arduous demonstrating required in the first term. All demonstrators before Christmas were staff members; all had been persuaded of the high priority of the first term's teaching; all had attended the training session before teaching started. Each demonstrator had his group of about ten students which he took through the parts of the first term's course in turn, he knew them all by name at the time of our visits, and he was the only demonstrator his students had encountered in spite of the irreplaceableness of each demonstrator entailed by this. At no time during the three-hour sessions did a demonstrator slip out for an odd five minutes to deal with an urgent matter elsewhere. All the four who demonstrated during our first visit remained for half an hour at the end of the session to assure us that these sacrifices were inseparable from the job and amply rewarded by its evident success.

We saw, at our first visit, the detailed way in which a three-hour session was spent. The students came promptly and went to the places where they had worked before. Each demonstrator gathered his group of students round a blackboard. One, apparently typically, spoke for fifteen minutes summarising what had been done last time, and then explained what was going to be done in the next half hour. No paper was distributed. He asked questions expertly phrased so as to receive answers of the kind he wanted. Then the group went back to their places to do the next thing, and the demonstrator went round asking how they were getting on and answering those who had questions. It was implicit in his manner that he had nothing else to do than attend to the problems that students could be expected to have, without embarrassing those who seemed for the time being not to have any. After less than an hour the group reassembled and the demonstrator asked for results and for explanations, which were both forthcoming. The demonstrator listened carefully to suggestions which individual students made, referred to a suggestion which a student had made earlier and which he had since thought more about, and asked the group to try out an idea which one of them had put forward. The demonstrator wrote up questions of a general nature, to be thought about at leisure, gave directions about the next stage of the work, and then, while the students got on with it, again went to them individually when they asked for help. Almost all the demonstrator's time was taken up like this. After about two hours there was a break when everyone went for tea. The tea was near the laboratory, and during it conversations between demonstrators were interrupted by individual students asking demonstrators about the work.

161

The pattern continued, and nobody left before the end.

For most experiments, students sit in a group, so that each can see what the others are doing, and conversation can be general. Many told us how they liked the sociable nature of this arrangement, in which they soon came to know one another.

Here is a major reason for success. Parallel with the case, argued in the article, for including the material taught in the first term, there has been contrived a set-up with a social appropriateness which goes unstressed. The first term is one spent among strangers in strange surroundings, lacking established habits and relaxations, with a wish to fit in with new people and new ways. There is anxiety about relationships and relative standing. The chance to sit around a table and learn whether others are finding everything just as interesting, perplexing, or tolerable, under the concealment of some simple task, would give satisfaction whatever the task.

Further, while all is still unfamiliar, simple tasks with a person there to explain them, encourage free communication between staff and students. The Liverpool staff do not dissipate these advantages by taking the role of assessors at a time when assessment would not augment motivation, and would inhibit free and easy relationships. The free communication is important for those times when the discussions, as these discussions are prone to do, swing suddenly from the trivial to the fundamental.

Could not the Liverpool first term as well be done by instruction by post-graduates, or by written programmes? Perhaps the Liverpool paper modestly underemphasises the staff's good sense in doing it exactly as they have. Students mentioned 'waiting to catch up': perhaps programmes done at their own pace would be the answer. But what of the secret fears of being the only one who does not see the point of it all? Why not use postgraduates? But how many of them could sustain forty eight hours of teaching with continuous development and no repetition? Might not the times when 'the fact that it's monotonous becomes more apparent' get out of hand?

The reason why so little of what we saw was tedious exercise is that the staff are convinced of its fascination. Nobody there alluded to any lack of enthusiasm, help, or involvement on the part of staff; indeed all acknowledged their debts to others. Significantly, the end of each year sees a general meeting to discuss improvements and changes for the coming year.

Valuable though the Liverpool experience is on the subject of teaching how to handle data, its greatest success would appear to lie in the much more hazardous area of motivating staff. And possibly that is all that successful innovation really needs.

8.4   UNIT LABORATORIES AT BIRMINGHAM

The electrical measurements laboratory discussed in chapter 6, section 6.4, is one, but not the only, form that a particular laboratory can take within the overall unit laboratory scheme. In the present section, the overall organisation is described, with examples of the possibilities it offers and the limitations it imposes.   About the scheme, Robert Whitworth writes:

In essence, the system of organisation of laboratory teaching developed in this department, and in use here for second and third year honours physics students since 1971, involves no more than dividing a large class into a number of groups and sending these groups one at a time to a sequence of 'unit laboratory' classes.   Each such unit has its own room, and is organised by two or three members of staff.   The students spend all their laboratory time in one unit laboratory for a period of several weeks and then move to another unit.   Each unit has its own subject area and an appropriate pattern of work.

This may sound a formal and rigid system of operation, but it was introduced precisely because of the opportunities it provides for diversity of activity and for greater interaction between staff and students.   The system does not guarantee either of these benefits, but it is a framework within which initiative can bring rapid and visible benefits to both staff and students.

THE SYSTEM FROM THE POINT OF VIEW OF STAFF

Each member of staff involved will belong to a small team of two or three people responsible for a particular unit laboratory, who have the responsibility of planning the activities of a group of students for a period ranging from five to eight weeks.   Their laboratory will have a specific subject, and will emphasise certain aspects of experimental physics.   For example, a laboratory on optics emphasises precise measurement and the analysis of data, while another on electrical measurements emphasises the selection of instruments and the design of experiments.   The balance between the emphasis of the various units is decided in consultation between staff responsible for them.

The staff of each unit decide how it is to be organised. There could be a small circus of experiments, a set programme of exercises, a project or some combination of these. Times can be set aside for demonstrations or class discussions, and at various stages the students can work individually or in groups. The staff may arrange that a group of students will be taught by one of them at a time, or they may arrange to share the teaching in various ways, as they consider desirable or practicable. Where close student-teacher relationships would be valuable, the first arrangement has advantages. In some units there are postgraduate demonstrators, but always with a member of staff in charge of each group.

The essential feature of the system is that each demonstrator has a definite responsibility for the course provided by his unit, and a personal responsibility to the students within it. If he cares to accept it as such, this is a challenge comparable with that of giving both a course of lectures and tutorials on them. It is a situation in which he can be stimulated to give of his best, and be rewarded by the response of students. It is a situation which allows the staff concerned to make changes whenever they feel it appropriate to do so. It is a situation which allows staff to choose amongst several patterns of working: the unit can be organised as a set programme of experiments; as a group activity; or (since students can be known individually) with work given according to individual needs.

THE SYSTEM FROM THE POINT OF VIEW OF STUDENTS

Each student belongs to a group of about twelve with which he stays during the year and will come to know well. The laboratory timetable for a second year student at Birmingham could look like this:

Transistor electronics	5 weeks
Optics	5 weeks
Electrical measurements	5 weeks
Project	12 weeks

(The project period involves pooling the resources of the unit labs, and is not run as a unit. However, it is included here as an essential component of the year's laboratory work.)

Each unit provides a different experience, but in every case there are ideas to understand, experiments to do, and various tasks to be completed in the fairly immediate future.

In each five-week unit an individual can expect the equivalent of three to four hours personal attention from a demonstrator. He is an individual whose progress will be known. His absence would be obvious. A well designed unit can offer him a stimulating and rewarding few weeks. Together, these features help to explain why almost all students are present and working most of the time, experience having shown that attendance at unit labs is extremely high, and unexplained absences very rare.

Hopefully at the end of the year every student will have encountered a balanced selection of experiments and techniques. He will have had experience of a variety of kinds of work such as designing circuits, selecting instruments, carefully collecting data, and computer analysis of results. The various units will have allowed him some choice, while covering for all students those areas staff feel to be essential.

In his third year a Birmingham student has a different time-table from that outlined above. He normally chooses two out of three eight-week units on solid-state physics, nuclear physics and technical physics. These are followed by a 'group study' (Black, Dyson and O'Connor 1968) which has some features in common with a unit laboratory although it is a full time activity and includes theoretical as well as experimental work.

ADVANTAGES AND DISADVANTAGES

The unit laboratory system was devised by a working party set up to consider ways of improving the conventional circus then in use, in which, although the actual experiments appeared to be quite good, there seemed to be a need to provide instruction in certain topics like the use of instruments and to provide sequences of experiments in areas like electronics. There was also the need to consider the introduction of project-type work and how best to prepare students to design their own experiments. Such activities do not fit easily into a circus where all experiments have to be in some sense equivalent, and where the selection done by any individual can be somewhat arbitrary. The unit system was devised as a possible solution to these problems, but it also appeared as a way of bringing a fresh stimulus to both staff and students. It was fortunate that the change-over was done at a time when finances permitted extensive modernisation of equipment in some of the newly created units.

A particular advantage of the system is one that at first sight appears contrary to its very nature. If organised appropriately it can shift the emphasis away from subject matter towards types of experimental activity. No longer does one only say that a student should do, for example, three experiments on a.c., two on optics, two on heat, and so on, but one can say in addition that everyone must do one piece of design work, one piece of construction, two pieces of elaborate data analysis, and even if desired spend one period of time amongst a research group. The subject title of a laboratory need be by no means its only, or even its most important, characteristic.

The change to unit laboratories unquestionably involves sacrifices, and has been opposed for this reason. A serious loss is in the range of experimental work available. The choice of the units and their contents has to be considered with care, but once it has been made the students do not have much opportunity outside their projects for doing experiments in other areas that may interest them. A circus can provide a wide range of opportunities, but the larger the range the less likely it is that any one experiment will be done by any particular student. The large circus does not guarantee that a wide range of topics are covered by any individual, but it does provide choice and the loss of such choice was initially one of the main objections of students to the scheme. A particular concern was the loss of the opportunity to choose a day off. This does not just indicate idleness; it can present genuine problems for both students and staff, but the output of work has been so much improved by keeping up the pressure of activity that it may be thought worth some loss of freedom.

From the staff point of view the units can bring a danger of monotony and a loss of opportunity for showing the unity of physics. In certain units this danger is more real than in others. It is important that staff are not required to teach in the same unit for too many years, and it is also important that the same staff who run the units also get involved in the running of the student projects, group studies, or other such activities.

SUMMARY

The unit laboratory system is a pattern of laboratory organisation by which a large department may gain some of the advantages of a smaller one. But it has advantages over a small department because from the larger numbers of staff available individuals can be found who will concentrate their attention on their own particular part of the laboratory course, putting effort and resources into developing it which can only be justified if it is to be used many times by different students during the year.

8.4.1 DISCUSSION

Visits to the Birmingham unit laboratories, and discussions with students about them, bring out at once how varied they are, and how various students differ in the value they place on each. A view that there is some special essential quality of a unit lab cannot survive such contact with them in operation.

Thus, in one laboratory a visitor might walk in on one day to a discussion about the planning of a forthcoming experiment, on another day see students working through simple exercises, and on a third day see them working in pairs on projects. In another laboratory, he might easily suppose himself to be in a conventional laboratory in a small physics department, except that all the experiments related to one topic. In a third, he might find students all working steadily through the same sequence of experiments from a bulky manual, with the demon-strator in charge standing waiting to be asked about any problems. Yet another (third year) unit laboratory is run by a man who has dispensed with all written instructions, preferring to discuss and explain each experiment directly with every student.

That is to say, the staff have exploited the chance the system gives them to run the laboratory in the way they think best fitted to its subject matter and aims.

Inevitably, staff have also used the opportunity to choose how much, or how little, involvement they feel able to accept. So one lab is run by individuals who are alone and permanently busy with a group of students, while in another staff leave students to get on alone, sharing between them the job of being present. The system permits involvement, but does not force it:

> 'The trouble was, we never had a demonstrator in there. We stood around - we just couldn't work out what was wrong with it. If he'd been there, we could have worked it out and got on with it. So we wasted a whole after-noon.'

Those staff who opt for greater involvement give themselves

the problem (common also to the Liverpool first term -see 8.3) that urgent demands on their time conflict sharply with their laboratory work. For example, one found himself unavoidably delayed for an hour on the very afternoon he had set aside to lead a discussion; work which no colleague could easily take over from him.

The existence of some unit laboratories firmly based on manuals and set experiments means that they at least can be staffed by people without any other close connection with the department's teaching, so tapping the pool of talent also tapped by many conventional laboratories. In both, the price can be a lack of concern for relevant parts of other teaching, so that it was possible to find a man in one laboratory thinking that students had not previously met an instrument used in that laboratory when in fact it was the subject of substantial and compulsory training in the previous year. At the same time, his teaching in that laboratory betrayed a clear grasp of its purposes.

Thus, one basic decision in designing a unit laboratory is whether to make it such as to be able to employ what in chapter 2 were called 'plug-in demonstrators', or whether to design it so as to demand, and exploit, special knowledge and skill which a replacement demonstrator would have to learn. Both have their advantages and disadvantages, and it is not surprising to find both in use. In general, though such a judgement is hard to quantify, the unit laboratories at Birmingham did seem to have made it harder than elsewhere for a visitor to find himself talking to a demonstrator who was doing no more than tolerantly completing his stint. It seems likely that this is related to the greater coherence of the job: running a laboratory sequence on a definite topic as opposed to being available for consultation about problems which might be of any nature and on almost any topic.

Students' views about the various unit laboratories are not simply related to their different structures. Sometimes personal preferences override all else, so that the same laboratory could be approved of by comparison with the first year by one,

> 'At least you talk to someone. That's what I didn't like about first year labs: you were given a bit of paper, you read it, and you did it. You didn't really think about it. At least here you had to think, see a demonstrator, ask a question.'

while another could take the opposite view for the same reason:

'In the first year you had the manual and you worked through it, and only if you were in desperate trouble, or wanted a thing marked, you'd see a demonstrator. In the unit labs, the demonstrator is walking around you all the time, and sees your mistakes all the time, and tells you what to do all the time, as if you're being led by the hand all the time.'

to which another replied:

'In the real world you're not expected to work on your own from a manual. You have to think for yourself in conjunction with other people. And that's what that unit lab is about. It requires some thought, and some physics, and some discussion.'

On the whole, students liked the definite focus and clear division of work into blocks:

'Before, you were on one experiment for four weeks, and then they would tell you that you had a lot left to do. In the unit lab you can look forward to the end - you can see an end.'

though some said that they got bored with five weeks on one topic, perhaps not knowing that in many other universities second year students spend a similar time on one experiment.

Judgements were influenced by features of many kinds. One lab, with a firm structure, was praised because,

'There was no feeling that you had to rush against everybody else, like there was in other unit labs, where you kept feeling you were falling behind.'

Some preferred a lab where they felt they had learned more that they could be sure of remembering, while others liked one on the grounds that the experiments were interesting and useful, despite serious deficiences they felt to exist in its organisation as compared to other unit laboratories. One laboratory was praised because it cleared up difficulties left over from a previous lecture course which was said to have been not too good.

The general point to be drawn from these varied reactions is that unit laboratories are like any other, dependent in their effect on many external and internal factors. The format guarantees little. There was however, a rather general feeling amongst students that the laboratory work in them was to be taken seriously, an impression confirmed by the high attendance rate. Complaints were specific and detailed, with rather little by way of apathy or of the mild rejection symptoms not unknown elsewhere.

Several students noted how they had come to know staff more personally. 'They' and 'us' still existed as clear divisions, but as ones which could perhaps be softened:

'It takes a while to break through and realise that they're human.'

REFERENCES

Black P J, Dyson N A, O'Connor D A (1968) 'Group studies', Physics Education, vol.3, no.6, page 289.

Black P J, Whitworth R W (1974) 'Unit laboratories in the department of physics at Birmingham University', in Studies in Laboratory Innovation, Group for Research and Innovation in Higher Education, The Nuffield Foundation.

Court G R, Donald R A, Fry J R (1976) 'Teaching practical physics I, II, Physics Education, vol.11, page 397 and page 488.

Richards M J (1974) 'A first year course in experimentation for mechanical engineers at Brunel University', in Studies in Laboratory Innovation, Group for Research and Innovation in Higher Education, The Nuffield Foundation.

9. Purposes and planning

9.1 THINKING ABOUT AIMS

'It is very hard to think about what you are doing in laboratory teaching.'

(staff demonstrator)

How shall I think about what practical work is for, and how might what I want to happen be made to happen? This is the puzzle around which this chapter revolves. In the rest of the book, we have attempted to clarify some ideas, expose some possibilities, and draw some conclusions, but we have also tried to show the complexity of all the issues, especially by frequent reference to actual examples. One way to bring matters to a rather clearer conclusion is to tackle the question of the aims of laboratories, and ways of realising those aims, less tangentially than has been done so far.

The reader who detects no absolute promise of unqualified recommendations and firm conclusions is correct. The fact is that people not only differ about what the aims of laboratory work should be, but also about how to look at them at all - about whether, for example, there is really any profit in trying to analyse them or specify them very precisely. Also, thinking about aims needs to proceed at different levels: the relevant questions and complexities are not identical for the man teaching day by day in the laboratory, for the man responsible for planning it, or for people arguing in departmental committees about the relative place of laboratory work of different kinds in the course as a whole. The issues need to be addressed from all these different points of view.

The chapter deliberately brings together aims and action, rather than discussing aims in the abstract, because aims make little sense divorced from what one intends to do about them, and because actions influence aims - doing one desirable thing may make it harder to do another, perhaps equally valuable thing, for example. We have tried to bring this out by beginning with a discussion of various ways of analysing aims, and attempting to show the limitations of such approaches, despite

171

their real virtues and attractions.

Any thinking about aims soon brings home that the underlying issues are about making choices. One cannot achieve everything one would like to achieve, not only because time is too short, but also because aims compete with one another. Nor is the laboratory divorced from the rest of university work, and its aims, while distinctive, interact with those of the rest.

A discussion which did not recognise these complexities would fall into triviality and lack of realism, but to admit complexity is not to deny the need to decide and to act. Somehow or other, priorities have to be got straight. So far as possible, therefore, we have tried to offer clear views about the issues involved, not as prescriptions, but as matters deserving of consideration. The difficulty of the task is neatly summed up by a remark which many demonstrators would echo, made by a student about practical work:

'I'm not quite sure how, but I'm sure it's good for you.'

9.2   ANALYSING AIMS

Thoughts about the purpose of practical work tend at once to take an overall standpoint, but at the same time to point to a number of definite purposes within that. Thus, one demonstrator said that its purpose was:

'to teach some experimental physics, and also to back up lectures. Teaching experimental physics - I mean things like looking out for the best way of doing something.'

and a student in the same department explained:

'Well eventually...we'll be doing practical work that nobody has ever done before. Then it will really count: the getting into good habits, of weighing up errors, getting the right techniques of measuring things, generally getting the feel for doing things in the right order and with the right methods.'

Such remarks offer a picture of an indissoluble general aim, which might be called 'learning the art of experimentation', which contains or implies a number of subsidiary aims, such as that mentioned by the demonstrator who said,

'We're all very keen on scientific writing, but we don't

do enough about it.'

and perhaps also some rather more ancillary aims, such as helping with lectures.

Attempts have been made to list common and important aims of this kind, partly so as to be able to pose questions about their relative importance. In a survey carried out in 1965, staff responsible for physics laboratories in all years, in thirty six departments of universities in the UK replied to a questionnaire asking them to rate eleven aims on a scale 1 - 10 of importance. (Chambers 1972). Recent graduates also rated the same aims. The list below shows the mean importances given to each aim.

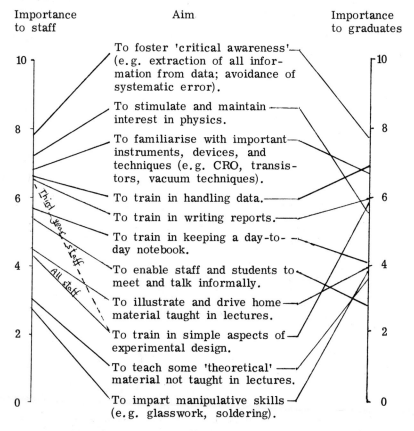

(Note that third year staff rate 'experimental design' about as highly as graduates, their rating being shown by the broken line.)

Interpreting these and other such lists of aims is difficult because they are of different kinds and at different levels. Thus 'maintaining interest' might be ranked high by those who thought it a precondition of anything else, and low by those who, while agreeing, thought it to be present in any case. Some aims can be taken as subsuming others: 'training in handling data' could be regarded as part of 'fostering critical awareness', for example. Opinions are also likely to be influenced by beliefs of practicality, so that 'teaching material not taught in lectures' might come low, not because people would not value it if they could have it, but because they do not think it possible to have it. The exact choice of words has some importance too: is 'train in simple aspects of experimental design' being read as meaning little more than kinematic design, or as devising an experiment to test a hypothesis?

Despite such reservations, which would arise in connection with any such study, the ratings of aims do give some picture of the balance of opinion, and as such bear comparison with the picture of practical work in action as portrayed elsewhere in the book.

Staff and graduates clearly agree in giving similar, fairly high ratings to the three aims of 'familiarising with equipment and techniques', 'training in handling data', and 'training in writing reports'; aims which all reflect definite recognisable parts of getting on with the practical job of working in a laboratory. Both place 'critical awareness' higher, and it is possible that they are seeing it as the nearest approximation in the list to the overall ability to experiment well; as what a student might have had in mind when he said:

> 'To start with, you're struggling to figure out things and jumping in headlong without thinking, whereas now you know from past experience that it's best to sit back for ten minutes and say to yourself, "Well, look, what's the best approach?", and you know that you can save yourself three or four hours if you sit down and think.'

Relatively, aims to do with teaching the ideas of physics are rated lower. One good reason may be that an aim of serving theory is often going to be in competition with others, especially those concerned with critical thinking. Examples of the choice that may need to be made here can be found in many of the laboratories described in the book, whether they emphasise one or the other. It must be hard for a student both to think critically about what he sees, and to regard it as a witnessing of the truth of what he has been told. The same experiment is rarely suited to both aims: the latter often implies some complexity of apparatus, and some cunning of design so that the point is clear, while the former is more likely to require simplicity, so that the student has a respectable chance of being able to think. At a more personal level, too, there is some competition. Many students have spoken to us of the laboratory as a welcome break from theory, so that a close link may be hard to bring off.

Chambers includes two aims, 'maintaining interest' and 'talking informally', which have more to do with how the laboratory might achieve its aims than with what its aims are. That the graduates rate both lower than staff, and rank informal talk last, could indicate that they have little trust in either coming about in the laboratory. Some of the reactions of students noted in chapter 3 are relevant here. The growing independence of students, and their wish to be left alone to sort things out themselves, can make it difficult to find the right way for staff and students to interact. That the fault, if it is a fault, lies in the structure of practical work and not in inherent qualities of staff or students, is suggested both by the strong and valued interaction widely mentioned in connection with projects, and, in the first year, by examples such as the reaction of students at Liverpool to the chance to talk easily and informally with staff.

That is to say, the specific, detailed, and individually challenging nature of the task set by a traditional experiment, taken together with the way the job of demonstrating is often conceived, can (but need not absolutely) militate against close contact. Not all would value it, but those who would might do well to consider steps such as working occasionally in groups; attaching demonstrators to students rather than to experiments and requiring them as part of the job to know students individually; reducing the scale by subdivision of the laboratory; or altering experiments so as to provide natural and necessary occasions for discussion.

It is not easy to reduce the gap between staff and students; a gap which is revealed by the kind of conversation in which a member of staff tactfully hints that students' knowledge is not what it might be, saying,

> 'I think you would do well in this experiment to have a thorough understanding of resonant circuits.'

only to be asked in return,

> 'What does "pico" mean?'

Such a distance in assumptions makes it hard, but even more important, to try for what one post-graduate described as:

'A friendly informal feeling which promotes cooperation and a good feeling about lab work.'

9.2.1 DIFFERENCES OF OPINION ABOUT AIMS

Broadly, except for the differences discussed above, staff and recent graduates agreed in Chambers' study. Where there are areas of disagreement between staff and students, it could be important to identify them, if only because matters often go better when all concerned agree about their purpose.

An example of such a study (Boud 1973) may be useful in suggesting how a department could itself conduct such an exercise internally, and obtain useful diagnostic information.

Boud asked first year undergraduates and staff in one university to rank aims of laboratory work, giving ranks both for what they took to be the aims of the actual course, and for what they would wish them to be in an ideal course. Results described below are intended only as examples of the kind of question such work might raise, not as things any other department than the one investigated might find.

When correlations between rankings were compared, it was clear that staff and students agreed rather closely on the aims of an ideal course, but not nearly so closely on the aims of the actual course, suggesting that what the staff thought the course was doing was not as evident at the bench as it might have been. Chapter 3, section 3.6, illustrates how such unintended messages can get across.

Information can be obtained from aims which are ranked very differently. In Boud's case students rated report writing and keeping a day-to-day notebook very highly as ideal aims, while staff gave them only a middling rank, much lower than, for example, developing skill at solving practical problems. Such a result cannot be interpreted without more evidence, but it could point to first year students' anxieties and lack of confidence more than to a judgement on their part that these things had special importance. He found (as did Chambers to a lesser extent) that students rated 'maintaining interest' much lower than did staff.

On differences between the actual and the ideal, staff felt that 'communicating technical ideas'was an aim that should be strengthened in an ideal course, while students thought it had too much prominence in the real one. Both, however, wanted 'teaching simple practical skills' to be more important than it was. Single such instances say little, but taken together, the differences and agreements could help to identify areas where, for example, staff may be running too far ahead of students, or where there is genuine disagreement which might well be brought more into the open.

The obvious difficulty in interpreting such data is that it is hard to know what an aim, at this level of generality, might mean. For the same reason, it is difficult to use such aims in actually deciding what to do: dozens of different experiments which commonsense says have important differences, can all plausibly be presented as serving some such aim as 'training in simple aspects of experimental design'. At the same time, their generality is inevitable: nobody could manage more than ten or twenty of them, in thinking about practical work, so that it is certain that they will for this reason not be very specific.

This is not to say that thinking at this level has no value. As already suggested, it can have use as a diagnostic tool. It can to some extent guide the balance and selection of work, especially in suggesting what to leave out (see, for example, a review (Aspden and Eardley 1974) of the design of practical work for Open University science courses).

9.2.2   TRYING TO SAY EXACTLY WHAT YOU INTEND

It suits the scientific temperament to imagine specifying the purpose of practical work rather exactly, and in terms of what could be observed, so as to be able to tell from observational data whether or not one has succeeded. Indeed, such attempts have been frequent in many areas of education (for a review of their value see MacDonald Ross 1973).

The writings of Nedelsky offer one example of an attempt at such an analysis for physics laboratory work (Nedelsky 1965, see also Nedelsky 1958). He proposes a set of objectives under the same three headings, 'knowledge', 'understanding', and 'ability to learn', which he uses for learning science in general. The three headings form some kind of hierarchy of increasing demand and scope. The list of objectives looks like this:

Laboratory knowledge :

Knowledge of apparatus and materials.

Knowledge of laboratory procedures.

Knowledge of relations between data and generalisations

from that data.

Laboratory understanding:

Understanding of processes of measurement:

Working of apparatus.

Method of measurement.

Understanding of experiment:

Experimental design.

Performing an experiment.

Interpretation of data.

Ability to learn from experiment or observation:

Ability to pursue experimental inquiry.

Possession of laboratory skills.

Disciplined thinking.

Imaginative thinking.

The immediate reaction to such a list is that it is vacuous. Without examples it conveys little, indeed it is not even clear if the form of the analysis is any good. Part of this reaction is due to not knowing the meaning which is being given to special terms. This is not to criticise Nedelsky, who does give plenty of examples of what he means, but to point out that arriving at any such list is to do more than to set down in an orderly way what most like-minded people agree upon. It involves ideas about learning (the progression from simply knowing about things through understanding ideas to using them to learn, for example), and beliefs about what is valuable (the fact that he includes 'imaginative thinking' tells us something about Nedelsky). All this is to say that proposing a set of objectives is not a cool technical job, but is to put one's ideas and beliefs on the table.

Nedelsky's examples of what he means take the form of instances of test questions that could be set to satisfy oneself that the objectives had been achieved, having made more specific what kind of task would be a suitable test. So, having explained that 'knowledge' is meant to concern familiar apparatus and methods, while 'understanding' is to be taken to mean describing and explaining something somewhat new, an example of knowledge of procedures is given:

'An ammeter-voltmeter circuit for measuring resistance is set up, not completely correctly. Inspect the setup, and write down a complete criticism of it: polarities, ranges of meters, size of resistances, connections, and circuit.'

It is interesting to notice how much even just one example conveys. Without it, the fact that Nedelsky is mainly concerned with first year introductory physics (in the American context) is not obvious; with the example that information is not surprising. More important, one forms a judgement of whether the words 'knowledge of procedures' mean something grandiose, something reasonable, or something rather trivial. That is to say, analysing aims has to get down to a pretty concrete level before an outsider understands what the analysis means.

One good feature of using a test of an objective to explain what it means is that the process induces realism and modesty. The gain in reducing the level of wishful thinking may be more important than any gain in precision.

Another good feature is that it may suggest new and useful teaching ideas. Thus, to say that one wants to develop understanding of what an experiment is, is one thing, but to do it is another. Nedelsky's answer to how to test for it is also his answer to how to begin to teach it: the 'experiment-predict-verify' pattern he uses for some tasks.

In these, the student is given apparatus, and is told that later (as a test) he will be asked to predict the value of some quantity - perhaps the distance a ball will roll along a carpet after rolling down a slope of an angle and from a height that will only be specified then. His job is to carry out preliminary experimentation which will give him the data needed to make any such prediction, with known accuracy. Then, in the test, he has to do it. The idea is not very different from that behind the course in experimentation devised by Richards (4.5).

It would be a mistake, however, to suppose that if Nedelsky's idea (or any other like it) is a good one, it is good because it accurately reflects a precise objective. Rather, it can be judged good if students and staff both find it sensible, interesting, and worthwhile. It is highly relevant that one student who had done a somewhat similar exercise said,

> 'I liked that better because there was a goal - to predict something.'

A difficulty with making objectives precise is that it is hard to avoid throwing the baby out with the bathwater. Those things one is clever enough to make precise are not always those which one thinks most important. An example from a different author (Flansburg 1972) may bear consideration. Wishing to make more exact the aim of 'being able to ask meaningful questions of nature', he first turns it into:

'Given a particular phenomenon, the student should be able to state a hypothesis whose acceptance or rejection will contribute to the scientific understanding of the phenomenon.'

Not satisfied, he then specifies what would count as an acceptable observable performance; in this instance the producing of at least one hypothesis implying a functional relationship between measurable quantities. In another instance, keeping adequate records is translated into the requirement that three out of four fellow students should be able to follow them.

Although Flansburg explains, tongue in cheek, that he would prefer something more subjective but accepts that at present the demand for measurable results from things as expensive as laboratory teaching may be irresistible, it is hard not to feel that here a steam hammer is cracking a nut.

Most teachers do not think about teaching in this detailed way, nor is it obvious that they are at fault. Material in chapter 3 shows how, very often, it is the immediate demands of an experiment, both in techniques and in ideas, which guide the thinking. A good case can be made out that the demands presented by a sufficient range of actual experiments will themselves amount to as much, or more, than the sum of carefully designed experiences based on some analysis of experimentation.

At the same time, by no means all experiments are 'actual' in the sense required by this argument. Indeed, all are artificial in some degree. It is possible to blind oneself to the fact that valuable and important things - making decisions perhaps - are in fact left out of most of one's laboratory experiments, by pretending that they are 'experiments' in every important sense of the word and so necessarily contain all that is valuable. At least, as is suggested by the account of unit laboratories (8.4), it may be a healthy byproduct of a little precise thinking about aims to shift attention from ensuring mainly that there is variety and balance in the subject matter of experiments, to ensuring that there is variety and balance in the different types of activity which they offer.

A workable and exact framework of analysis of aims of laboratory teaching, then, is hard to come by and examples of it do not invariably ring true. But if one makes no attempt at it, important things may escape attention. The laboratory is a place so much centred on specific jobs and their immediate problems, that it is hard for a teacher to deal on the spot with difficulties as they arise in such a way that what he does works always to a larger purpose.

9.3   THREE MAIN AIMS?

The foregoing discussion shows that analysing aims is difficult.
Chambers' commonsense approach has the problem that aims
then exist on several levels, and so are hard to sort out. But
the more radical approach of Nedelsky produces aims at a
necessarily high level of abstraction, which can only be made
understandable by immediate concrete examples. A crude but
reasonable new start might be made by noting that much practical work can be seen as serving one of three kinds of aim:

learning techniques

learning physics

learning the art of inquiry.

A particular example is offered by the first year at Bath
(6.2) where students are told that the purposes of the three
different kinds of experimental work are:

'...to teach you how to use certain basic measuring
instruments in a professional manner.'

'...to illustrate or amplify lecture material where
appropriate.'

'...to simulate the real life situation of the applied scientist
as closely as possible, though on a very small scale.'

Other examples would serve; the Bath case is clear because
different activities serve different purposes. In others, all aims
are, to a greater or lesser extent, involved in each experiment.
Even so, similar broad groups of aims are often discernible.
Two questions arise:

what, more exactly, is involved in each of these aims?

what case is there for providing distinct activities for each?

At first sight an analysis of what is involved in each of these
aims looks straightforward. Clearly techniques include the use
of instruments, the oscilloscope being the example that invariably
springs to mind. At later stages in physics, instruments like
phase-sensitive detectors may seem important. The use of a
computer, and learning to programme, is a case where most
people accept the sense of a distinct training course, whether
or not they take that view of learning to use, say, multimeters.
Similarly, workshop techniques are, where taught, often done in
this way. It seems plausible that where a technique uses special
facilities or is a substantial learning task, splitting it off seems
most natural, and 'picking it up en route' less so.

181

Where, however, does the list of techniques stop? Even a list of manual skills soon grows beyond reason if one sets about identifying all the skills students need. Worse, the list grows longer still when other important skills are taken into account. For example, what about the ability to communicate? The writing of technical reports can be, and often is, seen as a technique to be learned. What, however, about oral communication or working in a team? Few physics departments would be happy to be accused of producing graduates who could not get their ideas across, and who could not work with others. Just as few would be happy with any direct training in these areas (though see chapter 8, especially 8.2). Again, is the ability to plan and design an investigation usefully to be thought of as a set of techniques? If one wishes to argue that it is, will one go so far as to include critical thinking in the same category?

It becomes clear that the term 'technique' is not simply a neutral label. Rather, it marks off and distinguishes those things which are thought to be well suited to special training devoted directly and mainly to them alone. The question of what is well suited to such treatment remains open.

Turning to the second kind of aim, there is again a clear initial case for practical work designed at least in part to help students understand physics better. As students say:

'It's fine doing it in theory, but once you get in the lab and you're actually faced with a capacitor or something, it's not really the same.'

'It forms a concrete base for your learning, if you actually prove something in a simple form.'

'It helps you realise what you do and don't understand.'

This purpose of practical work is often called 'illustrating ideas taught in lectures', but that may be to misconceive it in two ways.

The first misconception is illustrated by the first quotation above: practical work is often not 'illustration', but learning at a new and different level. In the lecture, for example, the capacitor is two lines drawn on the blackboard, while between the plates is an electric field which is often the centre of attention; the main reality to be discussed. Then, in the laboratory, the plates can suddenly appal one by their gross physical presence, while between them there is just an absence; nothing at all. The electric field, so real on paper, turns back into a figment of the fertile imagination of theorists. Again, in lectures a harmonic oscillator seems to behave as it does because it satisfies a certain differential equation; in the laboratory one sees springs, masses, and damping vanes as causes.

The second misconception is that to think of experiment as 'illustrating' ideas demotes to an ancillary role what should be an equal partner in a dialogue. Physics is not that body of ideas which merely happens to come true when tried out: the interaction of theory and experiment is much more complex than that. Furthermore, it is that interaction which is physics, and which needs to be taught.

This, however, is to put the objections too strongly. There is a place for demonstrations, both in lectures and in some sense or other in the laboratory. There is a sense in which ideas seem to come alive, or gain substance, when they happen before one's eyes. It is not a matter of seeing being believing, so much as a matter of seeing more of what those beliefs are all about, and so becoming able to think about their likely application to actual instances more critically and more constructively. The real question, then, is about the role played by seeing or doing experiments in the process of learning ideas in physics, and in learning what it is to do physics.

What ideas need illustrating? As with techniques, the list is potentially enormous. Taken seriously, this aim gets out of control too. But it must be said that a situation in which lecturers rarely do demonstrations is not a satisfactory one, and that it is wrong to put upon the shoulders of the laboratory all the burden of illustrating ideas. Here the role of the laboratory interacts with the department's whole view of its teaching, with the effort people will put into demonstrations and with the support they get. The lecture should serve learning; the laboratory should not mainly be that servant's servant.

In the end, then, the discussion both of 'learning techniques' and of 'learning physics' as aims suggests that both aims overlap with and are part of the third kind of aim, 'learning the art of inquiry'. Doing and knowing are not ultimately separable from the act of working like a physicist.

When one asks what 'learning to inquire' might involve, the fact that inquiry is an art rather than a science makes any complete or rigorous analysis problematic. Examples of what is involved are not hard to give: they might include being able to plan ahead; being able to propose new solutions; extracting all possible information from data; taking seriously all possible objections; and so on.

Such a list is not, however, a closed one, nor are the items on it independent. For this reason, learning the art of inquiry is often thought of as the province of project work, with its combination of having to get on with the job together with the chance to learn alongside more expert practitioners, taking as much from what they do as from what they say.

The case for putting things together in some such way is hard to refute. Knowing diffraction theory is one thing, and being able to grind straight edges is another, but in designing and building slits for diffracting light the two are not compartmentalised: just how rough to leave the slits is a matter of theory and of workshop technology at one and the same time.

Looking back at the whole argument, its practical implication is that there are some choices to be made.

One choice is how far to go along with any analysis of laboratory learning into components. At one extreme, one might abandon the whole attempt. If one does not, the analysis still has to be cut off somewhere, as otherwise it grows beyond all reason. Some criteria for knowing when to stop are badly needed.

A second choice lies between providing special activities for particular components, or relying on a series of experiments each intended to develop bit by bit a large number of skills at different levels.

The choices are not absolute: a mixture of approaches may be as good as leaning strongly one way or the other. But they do still exist.

None of this will have much effect if the purpose of the work is not effectively communicated to students; if it does not seem useful and sensible at the time. So that is what the next section is about, before issues of choice are further explored.

9.4   TELLING THEM WHAT IT'S FOR

It is easy enough to write at the head of a script that the aim of an experiment is to measure some quantity. But then, as one staff demonstrator put it,

> 'The full potential of the old circus experiments is not realised. They think that the object is to find the inductance, not to learn how to find inductances.'

But it is hard to explain the overall purpose of a collection of experiments. An example which seems to have qualities of simplicity, directness, and intelligibility is the statement issued to second year students at Bristol. It includes:

'... you are given much less instruction about how to do the experiment, or even what experiment to do. You are expected to think these things out for yourself (with the aid of suitable references), a painful process very often, but one which forms a necessary introduction to the project work...'

... We should make it clear that we do not consider your lab work and your lecture work to have essentially different objectives. In your lectures you tend just to hear about idealized and isolated ideas, whereas in the lab work you have, more often, to connect the ideas together; that is, you practise physics in the round. When you are assessed, it will be essentially on your mental skills, not your manual dexterity, though this helps of course. Do not be surprised at this.

... marking sessions should be of the nature of tutorial sessions in physics: it should not surprise you if the thorough checking of an experiment takes an hour, or longer. The demonstrators will usually quiz you rather deeply on many aspects of the experiment. You should be prepared for such questioning, and be prepared to defend your ideas and your results. We want to know how well you have under-stood the experiment, how deeply you have thought about it, and how coherently you can talk about it. At the same time, if there are aspects of the experiment you do not understand, this is a good opportunity to clear them up.

... A good, readable prose style is valuable, especially for scientists who often have difficult ideas and subtle arguments to communicate. A good style is not obtained without effort and practice, but for our purposes an adequate workaday style can be obtained by asking oneself...

"What have I written? Is it what I meant to write, and was it worth writing?"

"Have I made it impossible for my meaning to be mis-understood?"

and, subject to these,

"Have I been as brief as possible?"

Your final reports are judged on the quality of presentation as well as on the quality of the physics...preferably being typed (there is a stark quality about typescript that forces one to think rather carefully about what one writes)...

...Students sometimes complain that demonstrators look for different things in reports...This is hardly surprising: they are only human, and science is a many-faceted thing, even in these relatively simple laboratory experiments...'

In many another laboratory, the discussion of aims is much shorter, more general in tone, and not conspicuous for giving much immediate practical advice. Typically, that is, one might read that the purpose is,

'...to illustrate important physical principles and the use of modern techniques of measurement.'

this being followed at once by details of laboratory hours and so forth. The Bristol notes are notable for tying aims and actions closely together; explaining what the laboratory is all for in part by explaining how things will or could be done. They have an air of realism which suggests that the writer means what he is saying. Recalling the previous discussion (9.2.2) about the difficulty of specifying objectives with precision and in terms of what could be observed, it is notable how well such notes might be thought to meet such criteria without taking the form of ind-ividual listed items.

Despite the virtues of saying what one means, much of chap-ter 3 illustrates how purposes are communicated at least as powerfully by practical experience; by what students find them-selves doing, and by how staff react.

This means that the organisation of the laboratory needs to have a certain simplicity and consistency. Chapter 5 shows how the pattern of work at Bristol does have something of this quality (5.4), as indeed it does elsewhere too. The point of the Bristol example is not to suggest that the choices made there are the best. The point is the rather good match of word and deed that it illustrates.

So, whatever kind of choice is made about how to look at aims, and about how to organise laboratory work so as to realise them, anyone running a laboratory would do well to ask himself if what he plans to happen has about it a clear self-consistency that delivers a clear message about his intentions. Another way to ask the same question is to ask whether it is possible to tell students plainly what will happen and why, in simple language. If not, there may be some matters worth attending to.

9.5   CHOICES AND CONSTRAINTS

What might be criteria by which to choose what kind of tasks to set in the laboratory, how to look at their aims, and how to organise them into an effective whole ? How will one choice affect others?

A choice to be made is the division of tasks between ones which are training in techniques of some kind, and ones which are more complete in the range of aims they tackle.

A first criterion for this choice might be convenience. It is convenient to set workshop practice apart, because it uses its own special resources. The same could apply to computer programming. It may be convenient to do it for a range of other things, such as learning to use complex instruments, to handle data, plot graphs, estimate errors, or write reports.

A second criterion might be emphasis. One disadvantage of the relatively undifferentiated experience offered by, say, a project, is that when many things are learned at once, some get lost because they get no conscious attention. So for example, even though the laboratory is plausibly the best place to develop good physical intuitions about such things as the likely size of an effect or times when simple theory cannot be applied to a situation, it less often happens than is wished. Students have very often had before their eyes things which would teach it, but have too often not seen them.

Emphasis is the reason for the arrangements at Liverpool (8.3). One of the experiments is contrived so that a whole group of students is led to expect that the period of a pendulum will not alter when the mass is changed, is made to discuss all their results until they have to admit that it does change, and is then led to think of reasons (such as that the heavier bob stretches the suspension more). By no stretch of the imagination could one suppose that one or two dramatised instances of this kind could teach much physical thinking, but at least they wave a flag for it:

> '...in a lot of experiments you've got to get the answer-
> then you get the answer and you relax. Now you really
> do look out - that answer could be way out.'

One real disadvantage of learning skills or techniques apart from their more 'real' application is that it can feel dull and mechanical. Steps, even artificial ones, may need to be taken to keep it interesting:

'In workshop practice we made brass cannons. X runs it - he's a real character. It doesn't matter what you make: it's the skills in the use of lathes, grinding machines, jewelling, and so on that matter.'

Things which are learnt apart are also rather too often not transferred to the situations where it is intended that they will be used. This sad psychological rule applies as much, of course, to knowledge gained in lectures, or to habits cultivated in tutorials, as it does to learning how to use instruments or calculate errors. Perhaps the best hope for effective transfer is when there are feelings of pride, pleasure, and achievement associated with the acquired knowledge or know-how. Encouraging that is more likely to be a matter of the enthusiasm and interest shown by staff, than of selecting the right subject matter.

Several examples in other parts of the book suggest that special bits of training work well when they are seen to have immediate point; when they are soon put to use. Where goals are long term, there has to be special care to make sure that the work seems sensible at the time. Training in report writing could be an example: paper exercises without any immediate pay-off in, say, helping with the writing of an actual report could be ineffective. By contrast, training in using new and interesting instruments with the immediate prospect of using them in a small project (6.4, 6.5) can work well.

Beyond these matters, two other criteria of choice might be simplicity and success.

Whatever pattern is chosen, there are advantages in simplicity, but simplicity has conflicting faces. A series of similar experiments of the classic kind has a simple pattern, but what is supposed to go on in each and be gained from each is far from simple. Students may find themselves quite unclear about what is important, about when to ask and when to go at a problem alone, about why the errors matter, and so on. Staff may not be clear as to what would constitute a good selection of experiments; about which ones might achieve which goals; about whether the omission of one could be balanced by the inclusion of another. A pattern of, say, skill training and demonstration experiments is simple in the sense that the parts serve fewer and better defined purposes. A clear articulation of different purposes through different activities has the advantage that it can be communicated to others - to post-graduate demonstrators for example - and that it itself communicates to students. At the same time, it would clearly be possible, and as clearly be foolish, to devise a complex system of individual bits of training of all kinds, which would lose the advantage of clarity of structure, no matter how well the objectives of each part were specified.

Perhaps the main trouble with simplicity is that the result is almost bound to look less impressive than one hopes. The question whether it is worth spending several hours on the CRO alone is bound to be asked. Against that, it may be necessary to remember that the actual work done on a series of measurements of physical quantities may be less impressive in scope and depth than it is often made to appear.

Lastly, the pattern of work chosen ought to be one which gives students non-trivial success, because succeeding is what more often leads people to try again than does failure. Thus there is something to be said for some rather definite small tasks near the beginning, to build confidence. Such a remark is not one which applies only to new first year students:

> 'It's always hard at the start, after the vacation. You take time to get back into things, and up to the standard they want.'

Where the choice is in favour of longer experiments throughout, it may not be wise to insist on a year-long circus of experiments of equal difficulty. Students change too much during the year for that pattern to be able to accommodate well to the change. An example of a simple but seemingly effective device is provided by the first year at University College London (5.2) in which the first term consists of a circus of shorter, easier experiments. By the second term, students uniformly spoke well of this gradual introduction. At Sussex, a different kind of success is encouraged by making sure that demonstrators know the experiments well, so that they can quickly and easily give practical help (5.3).

Other examples could be given. It may, however, be more useful for the reader to look again through the various accounts of laboratories, and of things which happen in laboratories, to decide for himself where they give students a real chance to do something well, to know it, and to be pleased about it.

9.6   TEACHERS AND STUDENTS IN THE LABORATORY

Education being a human enterprise, there is something lacking in a discussion which concentrates mainly, as we have done so far, on aims, organisation, and the nature of tasks. The best laid plans go astray if they do not suit those who have to put them into action. Plans which neglect the fact that teaching is done by people to people do so at their peril.

More positively, there is a part of learning to be a physicist which can not be learned from books, but only from working with physicists, and it is that part which the laboratory is well fitted to serve. Few would disagree, but few could deny the evidence that it happens too little. So, how can it be encouraged to happen more?

TEACHERS IN THE LABORATORY

In a sense, a demonstrator's main contribution is to be himself (2.3.1). A laboratory is more likely to succeed if it is organised so as to require what its demonstrators can best give.

In this respect, the traditional pattern has certain strengths. The immediate practical demands of the job are concrete and intelligible: helping with doing experiments. There is no untoward demand for commitment, so that laboratory teaching can fit in easily with other demands on time because one demonstrator can easily replace another. But a consequence of all this is that the job readily acquires a certain tedium.

The traditional pattern relies heavily on its aims being understood without having to be stated. That post-graduate students who have not themselves been through the laboratory have some difficulties is sufficient evidence of this. Staff, having more definite and individual views about what ought to go on, are in some danger of destroying the coherence of the laboratory by pursuing their individual lines, in the absence of any agreed policy. Against this, the traditional laboratory does allow, and even encourages, a valuable degree of variety of approach, so that a demonstrator can do what he is best at.

The unit laboratory pattern (6.4, 6.5, 8.4) and the tutorial pattern at Liverpool (8.3) offer a contrast. Here it is possible, and sometimes necessary, for the job of demonstrating to make greater teaching demands and offer greater satisfaction. In a laboratory such as the electrical measurements laboratory (6.4) the demonstrator can opt for the interest and pleasure of designing a whole laboratory and running it as he wishes, incorporating any kind of teaching, be it discussion in groups, demonstrations, or projects, that he sees fit and feels he can bring off. At Liverpool, the demands are more absolute, and such a laboratory can only succeed with the active cooperation and commitment of all concerned. Such participation is not easily to be had; it is no accident that policy in that laboratory is decided jointly in discussion by all the staff in it.

These more demanding patterns still exploit what the demonstrator has to give as himself, but ask him to give more of what he has it in him to give. The profit in better teaching and greater satisfaction must be offset against losses in flexibility. The more a person is individually involved, the harder he is to replace. Of some staff it would be sad to ask less; of others to make less unreasonable demands would be to get more.

People often work best when they feel that what they are doing belongs to them. To whom does the laboratory belong? In some the answer is nobody. In all, in different ways, the laboratory can be given an individual character, and where it is, that is a source of strength (amongst relatively traditional laboratories see several examples in chapter 5).

Much is often made of differences between post-graduate and staff demonstrators. Differences there are to be sure, but our visits to laboratories left no impression that they were sharp or absolute. Any job reserved for one in one place can be seen being done tolerably well by the other in another place. Where the job is difficult, such as leading a discussion or interviewing students closely about their work, experienced people often did it better, but not always. With help and advice, post-graduates could manage well, though organised help and advice was the exception rather than the rule.

Both parties, it seems reasonable to suggest, have much to learn from each other. Staff do often have a wider perspective, and can be less distracted by the details of an experiment, while post-graduates can bring a certain realism about how students feel and what they know. It could be a healthy step for any laboratory to introduce a regular discussion between staff and post-graduates, as normal routine. Such a discussion should perhaps begin with a run through the list of students, reporting progress and difficulties. Simply to expect everyone to know something about some students, as part of the job, would often help. If the discussion turned, as it might, to policy and organisation, so much the better.

That the laboratory has a greater permanence than its tenants, with directors changing every few years, and others more often still, is another reason for using discussion and consultation, as a means of making purposes more public. The change of people can bring in a healthy flow of new ideas, but this influx can be stifled by the apparent permanence and solidity of the laboratory, so that change is at best individual and piecemeal. A shared view of its purposes could assist continuous evolutionary change.

It must also be said, however, that there is value in individuality. No common view of desirable change ought to inhibit new and exciting ideas from being given a try.

STUDENTS IN THE LABORATORY

At one extreme, teachers rightly take pleasure in seeing a student taking off into serious experimental work of his own. His pleasure is the same as theirs in what is often called 'my own work'; his profit is similar too.

At another, staff look with satisfaction on the student who solidly gets on with the work he is set, trying to do what is wanted as well as possible.

Most practical work falls between these extremes, and is shadowed by uncertainties. Is this student independent, or just self-assertive and likely to turn in something meritricious? Is that one doing just what he is told, or just doing what he is told? Is the other not putting in ideas of his own because it is too soon to expect that, or because he thinks such action will be unwelcomed?

All these problems concern the interaction between what students are like and what they are asked to do. The cause of any particular outcome can be either, or both.

Demonstrators show some tendency to explain difficulties and successes in terms of students' characteristics. Indeed, there exist those who speak regretfully of how little of what they say seems to be taken to heart. With obvious sincerity, they explain that it appears that students do not have the ability to get anything like full value from practical work, with rare exceptions. Implied is the unquestioned correctness of the level of the work, and the laxness of providence in the supply of students.

Such a view finds it hard to survive visits to a variety of laboratories. There are differences which relate to the general quality of students, but there are equally large ones which do not. It is not hard to find students of no very remarkable standard doing projects in one place, at a time when others in another place of some reputation are carefully reined in for their own good. Within traditional experiments, the depth and rigour demanded varies both from place to place, and at different times in the same place. It is not that the quality of the end result varies notably, so much as that the means varies a good deal.

Something, then, of differences between what students do is left to be explained by differences in what they are asked to do. They tend stoutly to assert their independence in places where a reasonable degree of independence is assumed, especially when the expectation is confirmed by the local arrangements.

They tend to express gratitude for needed help in places where help is thoughtfully and carefully organised and given. They tend to persevere when given time to finish; to value 'finding themselves' when invited to pursue their own ideas; and to conquer difficulties when conquerable difficulties are put in their path. In a word, students like most people, live up or down to what is expected of them, if at all possible.

One lesson is to make sure that what is expected is what one values most. Another is that students are capable of more than most people have the opportunity to see them doing. But, before exploiting the principle that the more one expects the more one may hope to get, it is important to lay the ground carefully. Sudden new expectations, especially ones in conflict with those elsewhere in the system, can cause trouble.

Students are, necessarily, students. The trouble with this tautology is that, at the university, the student is hovering at the brink of adult independence. For part of the time, he gets on with the job of being a student: being talked at and trying to make notes and understand; attempting numerical problems; and doing a certain number of tasks in the laboratory. But at other times, whether often or not, he may delve into a subject just for its fascination, play with some calculation for some hours, or decide to look into some physical effect that interests him.

In the laboratory, this switching between work done for the staff and work done for private interest, between being a student and being a physicist, is not easy to handle. It comes out as experiments are chosen, as they are done, as they are written up, and as they are marked. At least a laboratory, however well organised to achieve valuable teaching aims, ought to permit and, if it can, encourage students to go into phenomena that strike them. A little slackness in the planning can have beneficial effects. It is not always evident that it is present in fact, though it is usually said to be, so a little organised slackness may be what is wanted.

The previous point extends to more than just letting students follow an interest. Part of the point of practical work is to find out how to do things for oneself, which both means being told, and means not being told. Students see this too:

'You've got a job in front of you...We've been told, "Do it this way", and obviously it works that way, but how can we develop our own system for doing things?'

The laboratory has an important contribution to make towards helping the student to find out about his own qualities and potentialities. A project, in particular, can help tell a student if he likes and is suited to research. That is worth having, but it is no less important to remember that many, even most, students will not do research, so that some flexibility in the nature of projects could be very desirable, as suggested in chapter 7.

Part of the question, 'What am I good at?' is the question, 'How do I feel about things?' So, when a student says,

> '...it gave me the first taste of doing something, think-about it...I actually rekindled my own interest in physics in four weeks...It really annoyed me that I couldn't think of reasons why certain things happened...Finally I got a picture of what was happening in my mind, and it gave me a kind of thrill really.'

one responds to how he feels about being like a scientist. These and other feelings are no less important than the know-how, knowledge, and competence that the laboratory may try to generate.

WORKING TOGETHER

One of the hardest puzzles to resolve about practical work is the balance between it as a private engagement with phenomena, and it as a place where people learn from one another. The balance that is in fact struck is often the only partially intended product of how the laboratory is organised; of whether students work singly or in pairs; of how demonstrators are organised and how they see their job; of what kind of tasks are set.

Visiting laboratories, the important virtues of students working alone, in pairs, or in groups were more striking than any major deficiencies of any one pattern. Good things could and did come out of all:

> 'My last long report took hours and hours - very deep. I chose to go fishing. It really interested me, so I decided to get as much out of it as possible.'

> 'I might suggest something, and he might say, "No, let's do it another way", and like that we get more ideas.'

> 'This idea of a group - helping each other. I find I've learned from it.'

This suggests using all three patterns, as they seem most

appropriate. Sometimes the task may be one better suited to working singly or in pairs, and this flexibility is worth having. Harder to accommodate is the likelihood that some students would at times be better off alone, or in company. To do anything about that would involve demonstrators knowing them well. Work done in groups can have special value, being best if the group has a particular job suited to that way of working. For example, one first year laboratory includes short projects done in groups of four, as a break in the pattern of experiments done singly. Clearly discussions (6.4, 8.2) and laboratory tutorial work (8.3) are well suited to group activity.

One of the more obscure conventions in many laboratories is whether students are supposed to help each other or not (3.3, 5.2). On balance, the realism and sense shown by students about what to ask for and what to give suggests encouraging rather than discouraging it.

Perhaps the greatest success of project work, after encouraging a sense of involvement, is to have found a natural way for staff to stop being 'demonstrators' and start being the useful neighbourhood expert. Learning directly from the man with green-fingered experimental expertise has long been part of the ideal of practical work, but realising the ideal has not usually proved easy.

One step in the required direction is to provide natural occasions for close contact. This can be done in many ways: by putting a man with a small group of students as in a unit laboratory; by removing all or most written information so that initial consultation is necessary; or by a tutorial format. With such arrangements, talk can turn if it wants to important issues.

Within the traditional framework, it is notable how some laboratories do produce at least some deep discussion and others do not. One thing that helps is to assign demonstrators to students, not to experiments. Another is to make routine a thorough discussion of work before it is finally written up. The latter is convenient to arrange where only some experiments are fully written up (5.4, for example). Alternatively, it could be valuable to introduce a mid-experiment consultation; a device sufficiently valuable in projects to be worth trying elsewhere. To have to explain and defend what he has done and plans to do, before it is too late, could be at least as useful to a student as to look back in retrospect on his mistakes.

At all events, one good test of a laboratory would seem to be the amount of useful discussion of physics between students and demonstrators. A laboratory director could do worse than take a session or two off listening in to what goes on, asking himself what limits it and when and why it takes off in a desirable direction.

At the same time, however, the laboratory is also a place for setting students free to get on with work that interests them. The key, perhaps, is responsibility. If the demonstrator always seems to want to know if the student is doing what he ought to be doing, compliance rather than responsibility is encouraged. If more often, he wants to know, and seems to expect the student to know, why he is doing what he is, freedom and critical discussion might naturally come into a better balance.

Finally, at the end of the day, it has to be said that few problems can be solved, and few desirable goals can be reached, in the absence of enjoyment, excitement, involvement, and interest. The inherent interest experiments may have can help, but much more important is the interest and pleasure shown by demonstrators in what students are doing. Anything which can aid that will do as much as the deepest thought or most careful planning to make practical work worthwhile.

9.7   PURPOSES AND PLANNING

This final chapter has been about the problems of planning to achieve various purposes. Its structure reflects the nature of the problems as we see them.

One approach has immediate appeal: first think clearly, listing what you want to achieve, and then set up kinds of practical work which will achieve these ends. Its appeal is that of simple, logical and rational design; its hidden implication is that traditional practical work is not clearly thought through, and so achieves good things, when it does, more by accident than by design. This simple view we reject, as it stands.

It is, however, good to think about aims. But, in identifying them, it is necessary to choose more than just a list of them. It is necessary to choose amongst ways of looking at them and amongst levels of thinking about them, from the abstract and global to the immediate and concrete. At these levels one is trading off meaning against perspective; the wider the point of view the harder it is to interpret or translate it in terms of action, while the more definite the point of view, the harder it is realistically to weigh up its general value. It is necessary also to choose when to stop analysing; to decide where to cut off the infinite regress into detail. When, say, a technique is to be seen as itself, or as an indivisible part of a wider and more far-reaching aim, is a matter of choice and judgement. All this is to say that thinking about aims is a matter of judgement based on beliefs. It is, and it always was; and no scheme of analysis can avoid the fact.

Setting out to achieve a set of aims is also trickier than it may look. Too little is usually said about the delicate balances and trade-offs which are involved. The book as a whole is in part intended to redress the balance. One of its messages is that it is important to watch out for hidden inconsistency, for conflict between aim and action. Such inconsistencies are to be found often where the reasons one can give students do not ring true (do the experiment as you judge fit, but be sure to do it properly). Aims then must be clearly expressed in events and in actions. Students are very good at reading events for the 'real meaning' of what is going on. Events, though, are tied to circumstances, and it is a practical necessity to choose what to do according to circumstances. To do that is a condition of achieving anything worthwhile in the real world, being at one and the same time hard-headed and idealistic.

Lastly, the single most important practical circumstance is the people involved. It is important to ask what they can do well. It is no less important to ask what might lead them to do better. Change of this kind can have as great an effect as any amount of improvement to hardware, to forms of instruction, or to administration, and without it the latter changes may have little effect. Ultimately, the most important resources of the laboratory are its human resources.

REFERENCES

Aspden P J, Eardley R (1974) 'Teaching practical physics: the Open University and other approaches' , occasional paper, Faculty of Science, The Open University.

Boud D J (1973) 'The laboratory aims questionnaire: a new method for course improvement?' , Higher Education vol. 2, pages 81-94.

Chambers R G (1972)' Laboratory teaching in the United Kingdom' in New Trends in Physics Teaching, vol. 2, UNESCO Paris 1972.

Flansburg L (1972) 'Teaching objectives for a liberal arts physics laboratory' , American Journal of Physics, vol. 40, pages 1607-1615.

MacDonald Ross M (1973) 'Behavioural objectives: a critical review', Instructional Science, vol. 2, pages 1-52. Elsevier.

Nedelsky L (1958) 'Introductory physics laboratory', American Journal of Physics, vol. 26, pages 51-59.

Nedelsky L (1965) Science Teaching and Testing, Harcourt Brace.

Index